T0309524

Viscous Flow

Cambridge Texts in Applied Mathematics

Viscous Flow

H. OCKENDON & J.R. OCKENDON
University of Oxford

CAMBRIDGE UNIVERSITY PRESS
Cambridge, New York, Melbourne, Madrid, Cape Town, Singapore,
São Paulo, Delhi, Dubai, Tokyo, Mexico City

Cambridge University Press
The Edinburgh Building, Cambridge CB2 8RU, UK

Published in the United States of America by Cambridge University Press, New York

www.cambridge.org
Information on this title: www.cambridge.org/9780521452441

First published 1995

A catalogue record for this publication is available from the British Library

Library of Congress Cataloguing in Publication Data

Ockendon, Hilary.
Viscous Flow / H. Ockendon, J.R. Ockendon.
p. cm. - (Cambridge texts in applied mathematics)
Includes bibliographical references and index.
ISBN 0-521-45244-9 - ISBN 0-521-45881-1 (pbk.)
1. Viscous flow. I. Ockendon, J.R. II. Title. III. Series.
QA929.025 1995
532'.0533-dc20 94-17066 CIP

ISBN 978-0-521-45244-1 Hardback
ISBN 978-0-521-45881-8 Paperback

Contents

Preface

These lecture notes are intended to provide third-year mathematics undergraduates who are already familiar with inviscid fluid dynamics with some of the basic facts about the modelling and analysis of viscous flows. Writing the notes has been an interesting task because so many of the phenomena to be described are not only associated with vitally important mechanisms in everyday life but they are also readily observable without any need for instrumentation. More sophisticated realisations are also readily available, for instance in the very valuable collection of photographs "An Album of Fluid Motion", edited by Van Dyke (2). Thus it is all the more stimulating when the mathematics that emerges when these phenomena are modelled is novel and suggestive of new methodologies.

The notes are strictly *not* self-contained and should be read in conjunction with standard texts which are referenced. We have concentrated on trying to present some of the salient physical ideas and mathematical ramifications as starkly as possible and, to this end, many of the exercises have been designed to be worked as an integral part of the notes; they are only put at the end of chapters for convenience. The starred exercises cover more advanced material and can be omitted at a first reading.

Experience has shown us that, in the twentieth century, theoretical mechanics generally has been one of the best vehicles for learning about physical applied mathematics. We hope that by showing students some of the basic theoretical framework which has developed as a result of the study of viscous flows, they will not only be able to delve further into the subject but also be well placed to exploit mathematical ideas throughout the whole of applied science.

We should like to thank our students Linda Cummings, Peter Howell, and Daniel Waterhouse, and our colleague Sam Howison for pointing out errors and ambiguities in earlier drafts of these notes. We should also like to thank Brenda Willoughby for typing the original manuscript and the many subsequent changes.

1

Modelling a viscous fluid

1.1 Modelling aspects

The principal motivation for the study of viscous fluid dynamics is the inability of the Euler equations of inviscid flow to predict certain familiar phenomena. We begin by reviewing these equations and some of their successes and failures.

In the simplest case of incompressible flow of an inviscid fluid, the Euler equations are

$$\left.\begin{aligned} \nabla.\mathbf{u} &= 0 \\ \text{and} \qquad \rho\left(\frac{\partial \mathbf{u}}{\partial t} + (\mathbf{u}.\nabla)\mathbf{u}\right) &= -\nabla p + \mathbf{F} \end{aligned}\right\} \tag{1.1}$$

for the velocity $\mathbf{u}$ and pressure p. Here the density ρ is constant and we assume that the body force $\mathbf{F}$ is prescribed. If $\mathbf{F}$ is conservative then these equations lead to the remarkable result that any flow that is initially irrotational will remain irrotational. This means that, for a wide range of flows of practical interest, $\mathbf{u} = \nabla\phi$ where ϕ is determined by the equation

$$\nabla^2\phi = 0 \tag{1.2}$$

together with suitable boundary conditions. This enables an enormous number of predictions to be made about inviscid flow which extend far beyond the simple examples found in textbooks on hydrodynamics. Some of the notable successes include free surface flows [Stoker], separated and bubbly flows [Birkhoff and Zarantonello] and aerodynamic flows (away from 'boundary layers' and 'wakes') [Milne-Thomson]. When compressibility is taken into account, by allowing ρ to vary and adding an equation of state to the model, even more phenomena can be described accurately. In particular the whole subject of acoustics is modelled in this way.

In spite of these successes there are many familiar situations which cannot be explained by equations (1.1) as illustrated by the following examples.

1. The first example comes from the classical theory of flight. It is possible to derive a theory of flow past aerofoils based on incompressible flow and using equations (1.1) but, unless we introduce an extra modelling assumption (the Kutta-Joukowski hypothesis), we encounter the D'Alembert paradox which precludes both lift and drag on either an obstacle placed in a uniform stream or an object travelling with constant velocity. The Kutta-Joukowski hypothesis, by introducing circulation around the object, does overcome this limitation but the origin of this circulation cannot be explained from the inviscid flow equations alone.

2. An everyday example is the observation that dust cannot be cleaned off the smooth surfaces of a car by simply driving fast. Inviscid theory for the airflow predicts a nonzero tangential component of velocity at a surface which the existence of the dust layer belies.

3. A common practical use for fluids is in lubricated bearings. The ability of a thin layer of fluid to support a large normal load while offering very little resistance to tangential motion is crucial in many kinds of machinery but cannot be explained by equations (1.1).

4. Another phenomenon observed in fluids is that they can be heated if external work is done on them. For compressible fluids, like the air in a bicycle pump, this can be explained by inviscid theory but for an incompressible inviscid fluid with boundaries held at constant temperature the model predicts that the temperature will remain constant throughout. This is in contradiction to experience in many situations; for example the oil in a bearing or fluid which is being injected under pressure into a thin mould can become significantly hotter without any change in density.

5. Another example comes from the theory of flight in the upper atmosphere. At altitudes between 10 and 100 km, the mean free path of a molecule in the atmosphere is so great that the macroscopic continuum model (1.1) is invalid. In this situation, as in some very small scale flows such as those involving free Brownian motion, some specific consideration must be given to the particle motion. This difficulty can only be dealt with via statistical mechanics and is beyond the scope of this book. However the idea of viscosity is relevant here too, as will be mentioned later in this chapter.

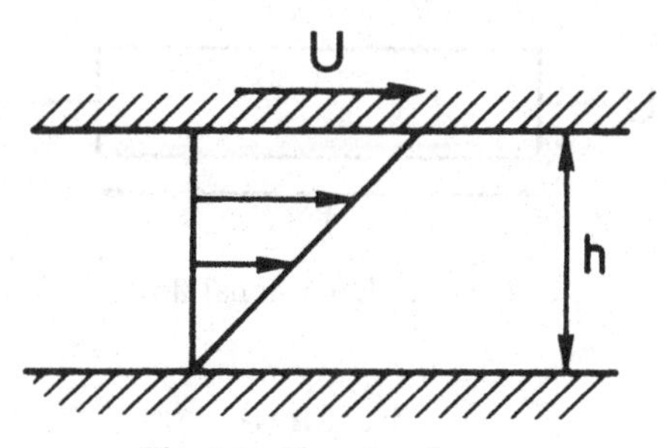

Fig. 1.1. Shearing flow

All these examples indicate the need for another model. The first four examples are all cases where *viscosity* or internal friction needs to be taken into account, and the aim of this book is to derive a model that will explain these phenomena. To do this we need to make just two basic experimental observations about how a viscous fluid reacts to shear forces and normal forces.

1.1.1 Shear forces

It is straightforward to set up an experiment to show that many fluids resist shear. In figure 1.1 the top plate is moved with speed U over a layer of viscous fluid of depth h lying on a fixed plate. The force required is found to be proportional to U/h, whereas for an inviscid fluid satisfying equations (1.1), the force required would be zero. A *viscometer* is a device which measures the viscosity of a fluid and is frequently based on this experiment. The constant of proportionality observed above is a direct measure of the viscosity of the fluid.

1.1.2 Normal forces

It is also easy to see that viscous fluids resist normal loading. In figure 1.2 the fluid (perhaps toffee) is pulled apart with velocity U and it is found that the required force is proportional to U/L.

These two simple observations will enable us to derive a model for viscous flow and this modelling is one of the most crucial aspects of the subject. The rest of this chapter will be devoted to modelling viscous flow, but at the end of this book (Chapter 5) we will consider briefly some other models of continua which can be thought of as generalisations of inviscid flows.

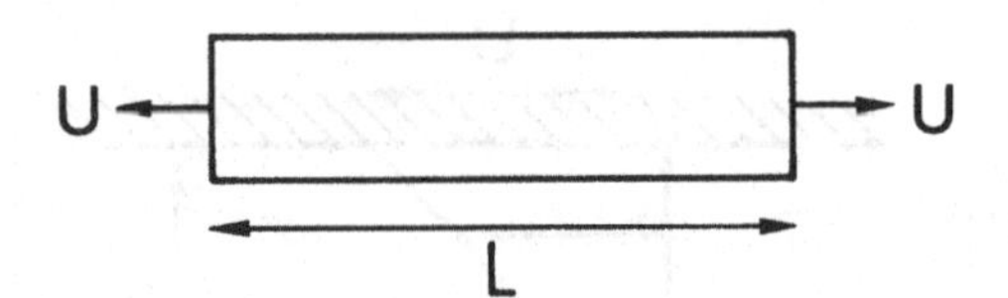

Fig. 1.2. Extensional flow

1.2 Stress

The first piece of evidence from the viscometry experiments in §1.1 is that the direction of the force exerted by the fluid on one side of a small surface element on the fluid on the other side of the element is not known a priori as it is for an inviscid flow. Therefore we need to define quantities σ_{ij} which are the force per unit area in the x_i direction[1] acting on a surface element whose normal is in the x_j direction. For an inviscid fluid, $\sigma_{ij} = -p\delta_{ij}$ where p is the pressure and δ_{ij} is the Kronecker delta defined by $\delta_{ij} = \{1 \text{ if } i = j, 0 \text{ if } i \neq j\}$. At first sight the introduction of σ_{ij} in place of the single variable p makes the modelling of a viscous flow a formidable task. However, we will see that we can work with just the nine quantities σ_{ij} referred to *one* specific set of axes and that, after certain assumptions, they can be written very simply in terms of derivatives of the velocity components. First, we need to show that, the 'stress' (i.e. force per unit area) on an *arbitrarily oriented* surface can be found in terms of these nine quantities by the application of Newton's law of motion to the tetrahedron of fluid shown in figure 1.3. If the force per unit area on triangle ABC is $\mathbf{F} = (\mathrm{F}_1, \mathrm{F}_2, \mathrm{F}_3)$ then

$$F_i \Delta ABC - \sigma_{i1}\Delta OBC - \sigma_{i2}\Delta OAC - \sigma_{i3}\Delta OAB =$$
$$\rho \, \mathrm{vol}\, OABC \times \text{acceleration in } x_i \text{ direction.}$$

If we now shrink the tetrahedron to zero and assume that the acceleration remains finite we obtain

$$F_i dS = \sigma_{i1} dS_1 + \sigma_{i2} dS_2 + \sigma_{i3} dS_3$$

where dS_1, dS_2, dS_3 are the areas of triangles OBC, OAC, OAB, respectively and dS is the area of triangle ABC. If $\mathbf{n}$ is the unit outward normal to ΔABC, $dS_i = n_i dS$ and so

$$F_i = \sigma_{i1}n_1 + \sigma_{i2}n_2 + \sigma_{i3}n_3 = \sigma_{ij}n_j \qquad (1.3)$$

[1] We will try to use the notation (x, y, z) for Cartesian coordinates throughout these notes but in this chapter suffices and the summation convention are unavoidable for reasons of space; hence $x = x_1$, $y = x_2$, $z = x_3$.

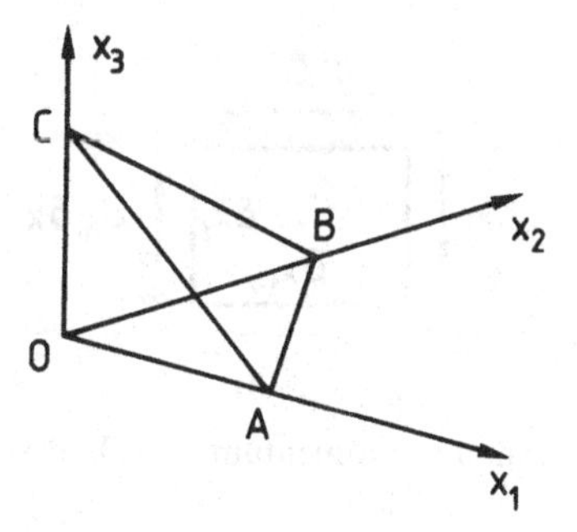

Fig. 1.3. Forces on a fluid tetrahedron

on using the summation convention. Thus we have found the stress on *any* surface element at *any* point in terms of the nine quantities σ_{ij}. These quantities form a 3×3 array, S, which, by using (1.3), can also be thought of as a matrix or linear transformation as follows.

We write (1.3) as $\mathbf{F} = S\mathbf{n}$ and let $T = \{t_{ij}\}$ be an orthogonal transformation which takes $\mathbf{x}$ into $\mathbf{x}' = T\mathbf{x}$. Then $\mathbf{F}$ and $\mathbf{n}$ will transform into $\mathbf{F}'$ and $\mathbf{n}'$ and (1.3) will be replaced by $\mathbf{F}' = S'\mathbf{n}'$. Substituting for these dashed variables in terms of the original ones and using (1.3) leads to

$$TS\mathbf{n} = S'T\mathbf{n}$$

and hence, since $\mathbf{n}$ is an arbitrary unit vector,

$$S' = TST^T \tag{1.4}$$

or, equivalently,

$$\sigma'_{ij} = t_{i\alpha}t_{j\beta}\sigma_{\alpha\beta}. \tag{1.5}$$

Thus σ_{ij} satisfies the usual rule for a linear transformation or matrix in a vector space (i = row, j = column). This is also the definition of a second rank *tensor*[2] and σ_{ij} is known as the *stress tensor.*

We can further simplify our modelling task by noting that we can reduce the nine quantities in σ_{ij} to six by consideration of the angular momentum of a fluid element. A general account is given in Batchelor [p.11] but for simplicity we shall just consider a rectangular element in two dimensions and apply Newton's law of conservation of angular momentum to this element. From figure 1.4, the rate of change of angular

[2] The same idea can be used to define a first rank tensor (or vector) which satisfies $a'_i = t_{i\alpha}a_\alpha$ and can easily be extended to tensors of third and higher rank.

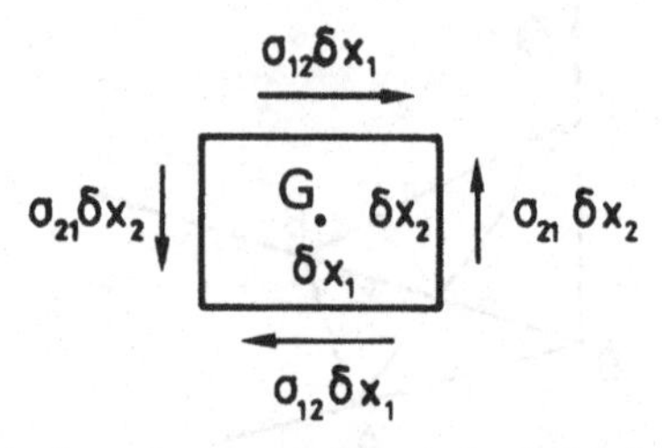

Fig. 1.4. Angular momentum of a fluid element

momentum about the centre of gravity G will be, to lowest order,

$$2(\sigma_{21}\delta x_2)\frac{\delta x_1}{2} - 2(\sigma_{12}\delta x_1)\frac{\delta x_2}{2}$$

where σ_{21} and σ_{12} are evaluated at G. Letting the rectangle shrink to zero and assuming the angular acceleration remains finite implies that

$$\sigma_{21} = \sigma_{12}.$$

This argument can be used in any plane and it follows that

$$\sigma_{ij} = \sigma_{ji}$$

and so the stress tensor is *symmetric.*

It is now convenient to separate σ_{ij} into two components: an *isotropic* part $-p\delta_{ij}$ as would exist in an inviscid fluid and a *deviatoric* part d_{ij} which is due to the viscous forces in the fluid. Thus we write

$$\sigma_{ij} = -p\delta_{ij} + d_{ij} \tag{1.6}$$

and we are now concerned with modelling the deviatoric stress. From the experimental observations mentioned in §1.1, we assert that d_{ij} varies linearly with the imposed velocity and inversely with the length scale of the apparatus. This leads us to infer that d_{ij} will be a linear function of the velocity gradients $\frac{\partial u_\alpha}{\partial x_\beta}$. This is the assumption that defines a *Newtonian Fluid.* It can be shown that $\left\{\frac{\partial u_\alpha}{\partial x_\beta}\right\}$ satisfies the linear transformation rule (1.5) and it is therefore a tensor (exercise 1). The assumption above implies that there exists a linear relation of the form

$$d_{ij} = A_{ij\alpha\beta}\frac{\partial u_\alpha}{\partial x_\beta}$$

between the deviatoric stress tensor and the tensor $\frac{\partial u_\alpha}{\partial x_\beta}$. Here $A_{ij\alpha\beta}$ is a tensor of rank 4 and, because there is no physically preferred direction in either d_{ij} or $\frac{\partial u_\alpha}{\partial x_\beta}$, it must be an *isotropic* tensor. That is to say it must have the same components in all sets of rotated Cartesian axes so

$$A_{ij\alpha\beta}t_{il}t_{jm}t_{\alpha\gamma}t_{\beta\delta} = A_{lm\gamma\delta} \tag{1.7}$$

for *all* orthogonal transformations t_{il}. The symmetry of σ_{ij} implies that

$$A_{ij\alpha\beta} = A_{ji\alpha\beta} \tag{1.8}$$

in addition, and we now need to determine the most general form of $A_{ij\alpha\beta}$ which will satisfy conditions (1.7) and (1.8). It can be shown by tensor methods that condition (1.7) on $A_{ij\alpha\beta}$ is sufficient to reduce the 81 scalar quantities in $A_{ij\alpha\beta}$ to just 3, and the symmetry condition (1.8) affords a further reduction to two so that

$$A_{ij\alpha\beta} = \lambda\delta_{ij}\delta_{\alpha\beta} + \mu\delta_{i\alpha}\delta_{j\beta} + \mu\delta_{i\beta}\delta_{j\alpha}$$

and, correspondingly

$$d_{ij} = \lambda\delta_{ij}\frac{\partial u_k}{\partial x_k} + \mu\left(\frac{\partial u_i}{\partial x_j} + \frac{\partial u_j}{\partial x_i}\right) \tag{1.9}$$

where λ, μ are scalar quantities.

We now show how this formula for d_{ij} can be deduced in a more straightforward way *without* resorting to the theory of tensors of rank 4. We first note that a fluid moving as a rigid body will experience no stress. Thus σ_{ij} will be zero when $\mathbf{u}$ is either a function of time alone or equal to $\boldsymbol{\omega} \wedge \mathbf{r}$ where $\boldsymbol{\omega}$ is a vector which varies only with time. This observation shows that σ_{ij} will depend only on the components $\frac{\partial u_1}{\partial x_1}$, $\frac{\partial u_2}{\partial x_2}$, $\frac{\partial u_3}{\partial x_3}$, $\frac{\partial u_1}{\partial x_2} + \frac{\partial u_2}{\partial x_1}$, $\frac{\partial u_2}{\partial x_3} + \frac{\partial u_3}{\partial x_2}$, and $\frac{\partial u_3}{\partial x_1} + \frac{\partial u_1}{\partial x_3}$. We therefore define the *rate of strain tensor*[3] $e_{ij} = \frac{1}{2}\left(\frac{\partial u_i}{\partial x_j} + \frac{\partial u_j}{\partial x_i}\right)$ and conclude that there exists a linear relation $d_{ij} = B_{ij\alpha\beta}e_{\alpha\beta}$ between the two *symmetric* tensors $d_{ij}, e_{\alpha\beta}$. This can be written more simply as

$$D = P_1e_{11} + P_2e_{22} + P_3e_{33} + Q_1e_{23} + Q_2e_{31} + Q_3e_{12} \tag{1.10}$$

where D is the matrix $[d_{ij}]$ and P_i, Q_i are symmetric 3×3 matrices. Thus there are now 36 unknown scalar quantities contained in P_i and Q_i. Now, rather than considering a general transformation, we consider a particular transformation T_1 which rotates the axes through $\frac{\pi}{2}$ about the x_1 axis. Then

$$\begin{pmatrix} x'_1 \\ x'_2 \\ x'_3 \end{pmatrix} = T_1\mathbf{x} = \begin{pmatrix} 1 & 0 & 0 \\ 0 & 0 & 1 \\ 0 & -1 & 0 \end{pmatrix}\mathbf{x} = \begin{pmatrix} x_1 \\ x_3 \\ -x_2 \end{pmatrix} \tag{1.11}$$

[3] Exercise 1 shows that e_{ij} is a tensor.

and the effect of the rotation on e_{ij} is to make

$$\left.\begin{aligned} e'_{11} = e_{11}, \quad e'_{22} = e_{33}, \quad e'_{33} = e_{22}, \\ \text{and} \qquad e'_{23} = -e_{23}, \quad e'_{31} = -e_{12}, \quad e'_{12} = e_{31}. \end{aligned}\right\} \tag{1.12}$$

The isotropy condition (1.7) means that

$$D' = T_1 D T_1^T, \tag{1.13}$$

so using (1.10) and (1.12) leads to

$$\begin{aligned} T_1P_1T_1^Te_{11} \quad &+ T_1P_2T_1^Te_{22} + T_1P_3T_1^Te_{33} \\ &+ T_1Q_1T_1^Te_{23} + T_1Q_2T_1^Te_{31} + T_1Q_3T_1^Te_{12} \\ &\equiv P_1e_{11} + P_2e_{33} + P_3e_{22} - Q_1e_{23} - Q_2e_{12} + Q_3e_{31}. \end{aligned} \tag{1.14}$$

Equating the coefficients of e_{11}, gives $P_1 = T_1P_1T_1^T$ and this shows (exercise 2) that $P_1 = \begin{pmatrix} \alpha & 0 & 0 \\ 0 & \beta & 0 \\ 0 & 0 & \beta \end{pmatrix}$ for some scalar α, β. By symmetry (or by considering rotations T_2, T_3 about x_2, x_3 axes) we can deduce that $P_2 = \begin{pmatrix} \beta & 0 & 0 \\ 0 & \alpha & 0 \\ 0 & 0 & \beta \end{pmatrix}$ and $P_3 = \begin{pmatrix} \beta & 0 & 0 \\ 0 & \beta & 0 \\ 0 & 0 & \alpha \end{pmatrix}$ and this makes the coefficients of e_{22} and e_{33} in (1.14) identical automatically. Similarly the coefficients of e_{23} in (1.14) are identical if $Q_1 = \begin{pmatrix} 0 & 0 & 0 \\ 0 & \gamma & \delta \\ 0 & \delta & -\gamma \end{pmatrix}$ and using symmetry and equating coefficients of e_{31} and e_{12} in (1.14) shows further that $\gamma = 0$. Thus we get

$$Q_1 = \begin{pmatrix} 0 & 0 & 0 \\ 0 & 0 & \delta \\ 0 & \delta & 0 \end{pmatrix}, \quad Q_2 = \begin{pmatrix} 0 & 0 & \delta \\ 0 & 0 & 0 \\ \delta & 0 & 0 \end{pmatrix}, \quad Q_3 = \begin{pmatrix} 0 & \delta & 0 \\ \delta & 0 & 0 \\ 0 & 0 & 0 \end{pmatrix}$$

and we have reduced (1.10) to a form which depends on only three scalar quantities.

We have so far considered only rotations through $\frac{\pi}{2}$ to achieve this great simplification but we can reduce the number of unknowns still further by considering a rotation through some other angle. We take $T_\theta = \begin{pmatrix} \cos\theta & \sin\theta & 0 \\ -\sin\theta & \cos\theta & 0 \\ 0 & 0 & 1 \end{pmatrix}$ and evaluate d'_{11} by the two expressions in

(1.13). From the left-hand side we get

$$\begin{aligned} d'_{11} &= (T_\theta D T_\theta^T)_{11} = d_{11}\cos^2\theta + 2d_{12}\sin\theta\cos\theta + d_{22}\sin^2\theta \\ &= (\alpha e_{11} + \beta e_{22} + \beta e_{33})\cos^2\theta \\ &\quad +2\delta e_{12}\sin\theta\cos\theta + (\beta e_{11} + \alpha e_{22} + \beta e_{33})\sin^2\theta \end{aligned}$$

and from the right-hand side

$$\begin{aligned} d'_{11} &= \alpha e'_{11} + \beta e'_{22} + \beta e'_{33} \\ &= \alpha(e_{11}\cos^2\theta + 2e_{12}\cos\theta\sin\theta + e_{22}\sin^2\theta) \\ &\quad +\beta(e_{11}\sin^2\theta - 2e_{12}\cos\theta\sin\theta + e_{22}\cos^2\theta) + \beta e_{33}. \end{aligned}$$

Finally, equating these two expressions for arbitrary θ leads to $\delta = \alpha - \beta$.

Thus the expression for d_{ij} can now be written in tensor form as

$$d_{ij} = \beta\delta_{ij}e_{kk} + \delta e_{ij}$$

and it can easily be checked that this is an isotropic tensor which transforms according to the rule (1.5) for *any* transformation t_{ij}. Finally, writing $\delta = 2\mu$ and $\beta = \lambda$ we recover (1.9). Therefore, having assumed a linear dependence of d_{ij} on $\frac{\partial u_\alpha}{\partial x_\beta}$, the most general expression for the stress tensor is

$$\sigma_{ij} = -p\delta_{ij} + \lambda\delta_{ij}\frac{\partial u_k}{\partial x_k} + \mu\left(\frac{\partial u_i}{\partial x_j} + \frac{\partial u_j}{\partial x_i}\right). \tag{1.15}$$

This is the constitutive equation for a *Newtonian Viscous Fluid* which depends on two scalar parameters μ, the *dynamic shear viscosity* and λ, the *bulk viscosity*. The former measures the response of the fluid to shearing and extension; λ measures the response to changes of volume and is irrelevant for the incompressible flows considered in these notes. These two quantities may depend on the local temperature, density, or pressure of the fluid but, in these notes, we will not allow them to depend on $\mathbf{u}$.

1.3 The Navier-Stokes equations

We are now in a position to apply the above ideas and formulate a definitive model for the motion of a viscous fluid. The only mathematical tool we need is the *Transport Theorem* which states that if $V(t)$ is a region which always contains the same fluid particles then

$$\frac{d}{dt}\left[\int\int\int_{V(t)} F(\mathbf{x},t)dv\right] = \int\int\int_{V(t)}\left(\frac{dF}{dt} + F\nabla.\mathbf{u}\right)dv \tag{1.16}$$

where the derivatives of F exist and $\frac{d}{dt} = \frac{\partial}{\partial t} + \mathbf{u}.\nabla$ is the convective derivative. This is a relatively simple example of differentiation under the integral sign (exercise 4).

1.3.1 Conservation of mass

By taking F as the density, ρ, in the above theorem (1.16), the principle of conservation of mass for any volume $V(t)$ leads immediately to the *continuity equation*

$$\frac{\partial \rho}{\partial t} + \nabla.(\rho \mathbf{u}) = 0, \tag{1.17}$$

on assuming suitable smoothness for $\rho, \mathbf{u}$ (e.g. differentiability in space and time). When ρ is constant, the fluid is said to be incompressible[4] and this equation reduces to

$$\nabla.\mathbf{u} = 0. \tag{1.18}$$

We shall assume that ρ is constant throughout the rest of this book unless specifically stated otherwise.

1.3.2 Conservation of momentum

Applying Newton's law to the fluid in $V(t)$, the rate of change of the momentum, $\int_V \rho \mathbf{u} dV$, equals the force exerted on the fluid in V. In the absence of body forces, like gravity or electromagnetic effects, the only forces acting on V are the viscous forces exerted on the boundary ∂V by the surrounding fluid. Using (1.3), the force in the x_i direction can be written as

$$\int\int_{\partial V} \sigma_{ij} n_j dS = \int\int\int_V \frac{\partial}{\partial x_j}(\sigma_{ij}) dV$$

by Green's theorem. If we now apply the transport theorem to this momentum balance and use equation (1.18) we see that

$$\rho \frac{du_i}{dt} = \frac{\partial}{\partial x_j}(\sigma_{ij}) \tag{1.19}$$

or, substituting for σ_{ij} from (1.15) and using (1.18) again,

$$\rho \frac{du_i}{dt} = -\frac{\partial p}{\partial x_i} + \frac{\partial}{\partial x_j}\left(\mu\left(\frac{\partial u_i}{\partial x_j} + \frac{\partial u_j}{\partial x_i}\right)\right). \tag{1.20}$$

[4] Note that some scientists, e.g., oceanographers, use $\frac{d\rho}{dt} = 0$ to imply incompressibility, but (1.18) still applies.

When the coefficient of dynamic viscosity, μ, is constant this equation simplifies further and can be written in vector form as

$$\rho \frac{d\mathbf{u}}{dt} = -\nabla p + \mu \nabla^2 \mathbf{u}. \tag{1.21}$$

It should be noted that $\nabla^2 \mathbf{u} =$ grad div$\mathbf{u}-$ curl curl $\mathbf{u}$ and that although $\nabla^2 \mathbf{u} = (\nabla^2 u_1, \nabla^2 u_2, \nabla^2 u_3)$ in cartesian coordinates it does *not* take such a simple form in polar coordinates (exercise 5).

Equations (1.18) and (1.21) are the *Navier-Stokes equations* for an incompressible Newtonian viscous fluid. All our subsequent work will be based on these exceedingly difficult equations. Before we begin to attempt to solve these equations we note the following points.

1. The equations are equivalent to the Euler equations when $\mu = 0$ but are of higher order when $\mu \neq 0$. This means that it is necessary to apply more boundary conditions than was possible when considering inviscid flow. We find that on a rigid impermeable boundary with normal $\mathbf{n}$ moving with velocity $\mathbf{U}$, it is now possible to impose the condition $\mathbf{u} = \mathbf{U}$, rather than $\mathbf{u.n} = \mathbf{U.n}$ which was all that could be used when solving the Euler equations. This at once removes our concern in §1.1 about the prediction of dust removal from a smooth surface.

2. We remarked earlier that a statistical or "kinetic theory" approach could have been used to model a gas as an ensemble of discrete particles. The key role is then played by a distribution function which denotes the probability of a particle being in a prescribed small region of phase space (position and velocity space). This function satisfies a famous integro-differential equation (called the Boltzmann equation [Chapman and Cowling]) and when we take the limit as the mean free path of the particle tends to zero we can retrieve (1.21) as a first approximation.

3. We have four equations for the four variables u_1, u_2, u_3, p. If the fluid is compressible ($\rho \neq$ constant) or if μ depends on the temperature T, we will need to consider, besides equations (1.18) and (1.20), some extra information in order to close the system. In any case, if we want to determine the temperature T, we need another equation and this is derived by considering the energy of the fluid.

1.3.3 Conservation of energy

We now appeal to the first law of thermodynamics which, loosely speaking, states that thermal and mechanical energy are interchangeable. (Depending on circumstances, other kinds of energy should also be taken into account; see exercise 6). In fact it is an everyday observation that mechanical energy can be converted into heat with great ease; none of the original mechanical energy ever seems to "leak out" of any practical system in the form of mechanical energy. However the reverse is far from true; indeed thermal energy inevitably leaks away to the surroundings when heat is being converted into mechanical energy.

It is in an attempt to put these observations into quantitative terms that the laws of thermodynamics have been formulated. However we can get a mathematical understanding by first writing down the statement of energy conservation inherent in the first law. The rate of change of total energy within $V(t)$ is

$$\frac{d}{dt}\int\int\int_{V(t)} [\rho c T + \frac{1}{2}\rho\,|\mathbf{u}|^2] dv \tag{1.22}$$

where c is the specific heat, which is assumed constant, and $cT = E$ is the internal energy per unit mass of the fluid. We now assert that this quantity must be balanced by the rate of working of the forces exerted on $V(t)$ by the surrounding fluid minus the rate at which heat is lost to the surrounding fluid. When we include heat conduction, with thermal conductivity k, but neglect effects such as radiation or chemical reactions, this may be written as

$$\int\int_{\partial V} \sigma_{ij} u_i n_j dS + \int\int_{\partial V} k \frac{\partial T}{\partial x_j} n_j dS$$

or, on using Green's theorem,

$$\int\int\int_V \frac{\partial}{\partial x_j}\left(\sigma_{ij} u_i + k\frac{\partial T}{\partial x_j}\right) dV. \tag{1.23}$$

We now use the Transport Theorem (1.16) and put expression (1.23) equal to (1.22). After using the Navier–Stokes equations the resulting equation can be reduced (see the incompressible version of exercise 7) to the equation

$$\rho c \frac{dT}{dt} = \nabla(k\nabla T) + \Phi \tag{1.24}$$

where $\Phi = \frac{1}{2}\mu\left(\frac{\partial u_i}{\partial x_j} + \frac{\partial u_j}{\partial x_i}\right)^2$. Note that the energy equation (1.24) uncouples from the mass and momentum equations when μ is constant. Since $\Phi \geq 0$, it gives the interesting prediction that almost any motion of a viscous fluid generates heat in a way that can be predicted precisely. More important, the fact that T satisfies a parabolic partial differential equation means that its evolution is certainly not reversible in time and this goes some way towards predicting the irreversibility which is observed in many real systems.

As an alternative to thinking in terms of partial differential equations, we can postulate the *second* law of thermodynamics. It is usually stated in terms of entropy which is an unnecessary concept for the purpose of these notes. We can get a good idea of its import from (1.24), which shows that in *any* viscous motion the heat of an isolated fluid can only increase; there is no way that a fluid can cool down simply by decreasing its strain rate. This irreversibility, which is absent in an ideal fluid ($\lambda = \mu = 0$), is associated with the fact that heat (or momentum) transfer can only occur between fluid particles when their temperatures (strain rates) are different and suggests that all viscous fluids, if left to themselves, return to an equilibrium state of rigid body motion. These statements are plausible deductions from our model under the assumptions that the relevant transport coefficients such as the viscosity, thermal conductivity, etc., are positive. Indeed, their positivity in this model is equivalent to a mathematical statement of the second law.

1.4 Vorticity

From equation (1.21) it is easy to show (exercise 10) that

$$\frac{d\boldsymbol{\omega}}{dt} = (\boldsymbol{\omega}.\nabla)\mathbf{u} + \nu\nabla^2\boldsymbol{\omega} \tag{1.25}$$

where $\boldsymbol{\omega} = \operatorname{curl} \mathbf{u}$ is the *vorticity* of the fluid and $\nu = \mu/\rho$ is the *kinematic viscosity*. This equation shows immediately that in two-dimensional flow, when $(\boldsymbol{\omega}.\nabla)\mathbf{u}$ vanishes, the effect of viscosity is to 'diffuse' vorticity.

An interesting concept related to vorticity is that of *helicity*, defined to be $\mathbf{u}.\text{curl}\mathbf{u}$. This quantity is especially important in the study of magnetohydrodynamics where the fluid velocity and the electromagnetic fields are coupled. It can be shown that the vanishing of the helicity is a necessary and sufficient condition for the existence of a family of surfaces that are everywhere orthogonal to $\mathbf{u}$; also, while an irrotational flow can be described just by one potential function ϕ, with $\mathbf{u} = \nabla\phi$, a zero-helicity flow can be described by two functions ϕ_1, ϕ_2 such that $\mathbf{u} = \phi_1\nabla\phi_2$ (exercise 10).

To understand the role played by viscosity in three dimensions, first consider the effect of the term $(\boldsymbol{\omega}.\nabla)\mathbf{u}$ in the inviscid case. Suppose we plot the trajectories of the two fluid particles which, at time t, are at $\mathbf{x}(t)$ and $\mathbf{x}(t) + \epsilon\boldsymbol{\omega}(\mathbf{x}(t), t)$ (**P** and **Q** in figure 1.5). After a small

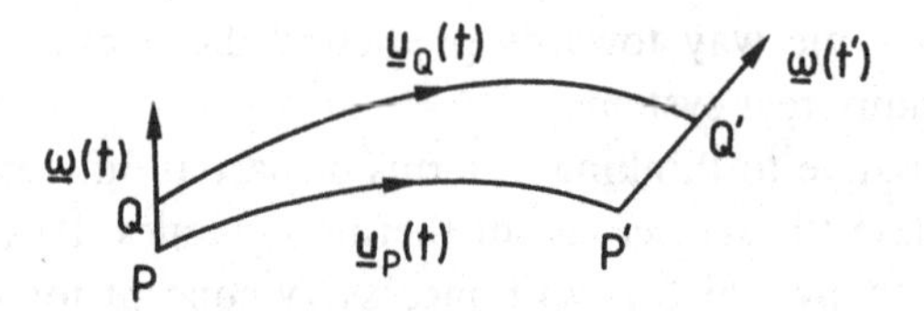

Fig. 1.5. Convection of vorticity

time interval δt, **P** will have moved to $\mathbf{P}' = \mathbf{x}(t) + u(\mathbf{x}, t)\delta t$ and **Q** to $\mathbf{Q}' = \mathbf{x}(t) + \epsilon\omega(\mathbf{x}(t), t) + u(\mathbf{x}(t) + \epsilon\omega(\mathbf{x}(t), t), t)\delta t$ so that by Taylor's theorem, $\mathbf{P'Q'} = \mathbf{PQ} + \epsilon(\omega.\nabla)\mathbf{u}\delta t$ to lowest order. But, from (1.25) with $\nu = 0$,

$$\epsilon\omega(\mathbf{x}(t+\delta t), t+\delta t) - \epsilon\omega(\mathbf{x}(t), t) = \epsilon(\omega.\nabla)\mathbf{u}\delta t$$

and so $\mathbf{P'Q'} = \epsilon\omega(\mathbf{x}(t+\delta t), t+\delta t)$ i.e. if **PQ** is proportional to ω at time t, **P′Q′** will be similarly related to ω at time $t + \delta t$. This argument can be extended to any time increment and it shows that vorticity is conserved in the sense of figure 1.5, rather than being convected with the fluid as it would be in two dimensions. If **P′Q′** is longer than **PQ**, the vorticity will increase as the vortex lines are "stretched" by the fluid. (This can be compared with the way an ice dancer increases her angular velocity by pulling in her arms.) However if viscosity is introduced this effect will be counteracted by vorticity diffusion. This smoothing effect of viscosity is especially important when the vorticity is concentrated in the neighbourhood of a 'vortex line' (e.g., a thin smoke ring). It can be shown that, in the absence of viscosity, such a vortex line will, when curved, move with an unbounded velocity (see Saffman). Hence viscosity, however small, is a crucial mechanism in vortex dynamics.

1.5 Nondimensionalisation

Before embarking on any mathematical analysis of a model, by far the most important procedure is the *nondimensionalisation* of all the variables. If this is done appropriately it allows us to estimate the relative sizes of the various terms in the equations and to judge whether any approximations can sensibly be made. Suppose, for simplicity, that there is just one speed U and one length scale L prescribed in the problem. Since there are also at least two other dimensional parameters, ρ and μ, involved, there is no unique way of nondimensionalising, but a natural choice would be to write $\mathbf{x} = L\mathbf{x}'$, $\mathbf{u} = U\mathbf{u}'$, $t = \frac{L}{U}t'$. It is less obvious what scaling to use for the pressure but, for the moment, we choose $p = \rho U^2 p'$ and substitute into (1.21) to convert the equation into nondimensional form in terms of the dimensionless variables $\mathbf{x}', \mathbf{u}', t', p'$. In order to avoid cumbersome notation we immediately drop the primes and try to remember that we are now working in dimensionless variables. Then equations (1.18) and (1.21) become

$$\frac{d\mathbf{u}}{dt} = -\nabla p + \frac{1}{Re}\nabla^2\mathbf{u}, \quad \nabla.\mathbf{u} = 0 \tag{1.26}$$

where $Re = \rho\frac{UL}{\mu}$ is the *Reynolds number*. For air and water, the kinematic viscosity, $\frac{\mu}{\rho}$ or ν, is 1.5×10^{-5} and 1.1×10^{-6} respectively in s.i. units[5] and thus for many everyday observations of these fluids, $Re \gg 1$. It must be remembered however that the quantities U and L were introduced from the boundary conditions of the problem and we should be careful not to assume that they will be the correct scales throughout the flow since the fluid may generate its own scales in response to these boundary conditions; indeed a glance at a plume of cigarette smoke shows that this can easily happen. Also we have arbitrarily scaled p with ρU^2 which turns out to be correct when $Re \gg 1$. However if $Re \ll 1$ then the viscous term $\nabla^2\mathbf{u}$ is multiplied by a large parameter and appears to dominate equation (1.26) and we will consider this regime in Chapter 3.

This completes the preliminary modelling stage. We have obtained the equations of motion for a viscous fluid but, because they are nonlinear, there are relatively few exact solutions and these solutions only apply to very specialized situations. Some of these exact solutions may be found in exercises 11-15 or various books [e.g., Acheson, Ch.2], but in these notes we shall concentrate on trying to get solutions which are useful in practical situations. This means that we shall have to consider approximations and, in particular, we shall exploit the fact that the Reynolds number is often either large or small. Before considering such solutions of the Navier-Stokes equations we need to develop some theory for approximate solutions of differential equations which contain a small parameter and we shall do this as we need it. In the next chapter we consider flows for which $Re \gg 1$ and we start by considering the theory of asymptotic solutions of differential equations in which the coefficient of the *highest* derivative is a small parameter.

[5] ν = acres/annum is a useful mnemonic for those with bad memories for dimensions!

Exercises

1 Show that $\frac{\partial u_i}{\partial x_j}$ is a tensor of rank 2, and that $e_{ij} = \frac{1}{2}\left(\frac{\partial u_i}{\partial x_j} + \frac{\partial u_j}{\partial x_i}\right)$ is a symmetric tensor of rank 2.

2 The orthogonal transformation $T_1 = \begin{pmatrix} 1 & 0 & 0 \\ 0 & 0 & 1 \\ 0 & -1 & 0 \end{pmatrix}$ and P_1, Q_1 are symmetric 3×3 matrices. Show that if $P_1 = T_1 P_1 T_1^T$ and $Q_1 = -T_1 Q_1 T_1^T$ then P_1 is a diagonal matrix of the form $\begin{pmatrix} \alpha & 0 & 0 \\ 0 & \beta & 0 \\ 0 & 0 & \beta \end{pmatrix}$ and Q_1 is of the form $\begin{pmatrix} 0 & 0 & 0 \\ 0 & \gamma & \delta \\ 0 & \delta & -\gamma \end{pmatrix}$.

3 In *two-dimensional* flow assume that $d_{ij} = B_{ij\alpha\beta} e_{\alpha\beta}$ where i, j, α, β take the values 1, 2. By considering a general two-dimensional rotation $T = \begin{pmatrix} \cos\theta & \sin\theta \\ -\sin\theta & \cos\theta \end{pmatrix}$, and using the tensor properties of d_{ij} and $e_{\alpha\beta}$, show that $B_{ij\alpha\beta}$ depends on three parameters. Show that by *further* considering $T = \begin{pmatrix} 1 & 0 \\ 0 & -1 \end{pmatrix}$, it is possible to eliminate one of the parameters and hence that $d_{ij} = \lambda\delta_{ij}e_{kk} + 2\mu e_{ij}$. What is the geometrical significance of $T = \begin{pmatrix} 1 & 0 \\ 0 & -1 \end{pmatrix}$?

4 Show that if $V(t)$ is a volume which contains the same fluid particles for all time then

$$\frac{d}{dt}\int\!\!\int\!\!\int_{V(t)} F\,dV = \int\!\!\int\!\!\int_{V(t)} \left(\frac{\partial F}{\partial t} + \mathbf{u}.\nabla F + F\nabla.\mathbf{u}\right) dV$$

for any differentiable function $F(\mathbf{x}, t)$.

5 Show that $\nabla^2\mathbf{u} =$ graddiv$\mathbf{u}-$ curlcurl$\mathbf{u}$ and that, if $\mathbf{u} = u(r, \theta, \phi)\mathbf{e}_r$ in spherical, polar coordinates, then

$$\nabla^2\mathbf{u} = (\nabla^2 u - \frac{2u}{r^2})\mathbf{e}_r + \left(\frac{2}{r^2}\frac{\partial u}{\partial\theta}\right)\mathbf{e}_\theta + \left(\frac{2}{r^2\sin\theta}\frac{\partial u}{\partial\phi}\right)\mathbf{e}_\phi.$$

6 The *first law of thermodynamics* states that the mechanical work done externally on a system together with any other externally supplied energy (e.g., heat) equals the change in internal energy E of the system. This implies that mechanical and thermal energies (and also other "internal" energies such as electromagnetic and

chemical energies) are interchangeable with each other. Show that the energy equation for a volume V of fluid can be written

$$\frac{d}{dt}\int\int\int_V E dV = \int\int_{\partial V} \sigma_{ij} n_j u_i dS - \sum_m \int\int_{\partial V} \mathbf{q}_m.\mathbf{n} dS + \sum_m \int\int\int_V R_m dV$$

where $\mathbf{q}_m$ is the outward flux of energy of type m across the boundary ∂V and R_m is any volumetrically supplied energy of this type. List the experimentally observed "constitutive" relations for $\mathbf{q}_m$, R_m for as many types of energy as you can (e.g. for heat conduction $\mathbf{q}_m = -k\nabla T$ and R_m could be microwave radiation).

7 Assuming the transport theorem (exercise 4) and the Navier-Stokes equations (1.21) for a viscous conducting fluid with velocity $\mathbf{u}$ and temperature T, show from the formula of exercise 6 that the energy equation for a compressible fluid may be written

$$\rho c\left(\frac{\partial T}{\partial t} + u_i \frac{\partial T}{\partial x_i}\right) = k\nabla^2 T + \Phi - p\nabla.\mathbf{u}$$

where the dissipation $\Phi = \lambda\left(\frac{\partial u_i}{\partial x_i}\right)^2 + \frac{1}{2}\mu\left(\frac{\partial u_i}{\partial x_j} + \frac{\partial u_j}{\partial x_i}\right)^2$ when the specific heat c and conductivity k are assumed constant.

8 Show that for an incompressible fluid, the dissipation Φ is zero if and only if the fluid is in rigid body motion.

*9 For a compressible fluid $\left(\frac{\partial u_i}{\partial x_i} \neq 0\right)$ show that

$$\frac{1}{3}\sigma_{ii} = -p + (\lambda + \frac{2}{3}\mu)\nabla.\mathbf{u}$$

and

$$\begin{aligned}\Phi &= \frac{2}{3}\mu\left((e_{11} - e_{22})^2 + (e_{22} - e_{33})^2 + (e_{33} - e_{11})^2\right) \\ &\quad + 4\mu(e_{12}^2 + e_{23}^2 + e_{31}^2) + (\lambda + \frac{2}{3}\mu)(\nabla\mathbf{u})^2.\end{aligned}$$

The kinetic theory of monatomic gases suggests that the quantity $\lambda + \frac{2}{3}\mu$ is zero (Stokes condition). Show that if $\lambda + \frac{2}{3}\mu = 0$ the dissipation is zero not just in rigid body motion but also for a spherically symmetric contraction or expansion.

10 (i) Show that the momentum equation (1.21) for an incompressible fluid can be written as

$$\frac{\partial \mathbf{u}}{\partial t} - \mathbf{u} \wedge \boldsymbol{\omega} + \nabla H = -\nu \nabla \wedge \boldsymbol{\omega}$$

where $H = \frac{p}{\rho} + \frac{1}{2}|\mathbf{u}|^2$ and hence show that

$$\frac{d\boldsymbol{\omega}}{dt} = (\boldsymbol{\omega}.\nabla)\mathbf{u} + \nu\nabla^2\boldsymbol{\omega}.$$

(ii) If $\Gamma(t) = \oint_{C(t)} \mathbf{u}.d\mathbf{s}$ is the circulation round a closed contour $C(t)$ which is moving with the fluid, derive Kelvin's result

$$\frac{d\Gamma}{dt} = -\nu \oint_C (\nabla \wedge \boldsymbol{\omega}).d\mathbf{s}.$$

*(iii) It can be shown that an arbitrary vector field $\mathbf{u}$ can be written as $\mathbf{u} = f\nabla g + \nabla h$, where f, g, h are called Clebsch potentials. Show that $h = 0$ is both necessary and sufficient for $\mathbf{u}$ to have zero helicity[6]. (Consider the integrability of $udx + vdy + wdz$ where $\mathbf{u} = (u, v, w)$).

11 Two layers of viscous, incompressible, and immiscible fluid flow in the x-direction under a constant pressure gradient between parallel fixed planes at $y = \pm a$. Each layer has a thickness a, both fluids have density ρ and the viscosities are ν_1 and ν_2.

What are the boundary conditions at $y = \pm a$ and $y = 0$?

Use the Navier-Stokes equations to find the fluid velocities in terms of the constant pressure gradient $\frac{dp}{dx}$.

If the volume fluxes of the two fluids are F_1, F_2 respectively, show that

$$\frac{F_1}{F_2} = \frac{\nu_2(7\nu_1 + \nu_2)}{\nu_1(7\nu_2 + \nu_1)}.$$

12 A viscous fluid is initially at rest between two infinite planes at $y = 0, h$. At time $t = 0$ the plane $y = h$ is set in motion impulsively so that it has a constant velocity U in the x-direction for $t > 0$. Show that after the impulse the velocity in the fluid is

$$\frac{Uy}{h} + \sum_{n=1}^{\infty} \frac{2U}{n\pi}(-1)^n \sin\frac{n\pi y}{h} e^{-\frac{n^2\pi^2}{h^2}\nu t},$$

where ν is the kinematic viscosity of the fluid.

Hence show that the stress per unit area on the plane $y =$

[6] The authors are grateful to Dr P.R. Baldwin for drawing their attention to this point.

h decreases monotonically in t from the initial impulse to an ultimate value of $\frac{U\mu}{h}$.

13 A fluid flows steadily along a straight cylindrical pipe of uniform cross-section D. The x-axis is along the pipe and the velocity in the x-direction is u. Show that u satisfies the equation

$$\mu\left(\frac{\partial^2 u}{\partial y^2}+\frac{\partial^2 u}{\partial z^2}\right)=c \quad \text{in} \quad D$$

with $u=0$ on the boundary ∂D, where c is the constant pressure gradient in the x-direction. Show that the drag per unit length on the pipe wall is c times the area of cross-section. Show that if D is a circle of radius a (Poiseuille flow), the total flux down the pipe is $-\frac{\pi c a^4}{8\mu}$.

If the cross-section of the pipe is an equilateral triangle bounded by $y=0$, $y \pm \sqrt{3}z = \sqrt{3}a$, show that an explicit polynomial solution exists for u. Show further that it is only possible to find such a solution for a cross-section bounded by a closed rectilinear curve in the case when that curve is an equilateral triangle.

14 Incompressible fluid occupies $y > 0$ above a plane rigid boundary $y=0$ which oscillates to and fro in the x-direction with velocity $U\cos\omega t$. Show that $\mathbf{u}=(u(y,t),0,0)$ satisfies

$$\frac{\partial u}{\partial t}=\nu\frac{\partial^2 u}{\partial y^2}$$

and, by writing $u = Rl(f(y)e^{i\omega t})$, show that

$$u = Ue^{-\sqrt{\frac{\omega}{2\nu}}y}\cos\left(\sqrt{\frac{\omega}{2\nu}}y-\omega t\right).$$

What is the width of the 'boundary layer', i.e., the order of magnitude of y for which the velocity is not "exponentially small" as $\nu \to 0$?

15 An incompressible fluid of viscosity μ fills the space between two concentric infinitely long cylinders of radii a and $b (a < b)$. The inner cylinder is fixed and the outer cylinder moves with a screw motion such that it has angular velocity ω about the common axis and velocity u parallel to that axis. The flow is steady and there is no pressure gradient. Show that the azimuthal velocity

component, u_θ, is given by

$$\frac{b^2\omega(r^2 - a^2)}{r(b^2 - a^2)},$$

where r is the radial distance from the axis, and find the axial velocity component u_z.

Show further that the maximum value of the pressure is attained on the outer cylinder.

*16 Assume that when a glass fibre is being manufactured by being pulled at each end $x = \pm Vt$, its velocity $u(x, t)$ is unidirectional and its cross-sectional area is $A(x, t)$. Show that conservation of mass then implies that, to lowest order,

$$\frac{\partial A}{\partial t} + \frac{\partial}{\partial x}(Au) = 0.$$

Assuming that the "extensional" stress σ_{11} is proportional[7] to $\mu\frac{\partial u}{\partial x}$ show that, when the inertia of the glass is neglected,

$$\frac{\partial}{\partial x}\left(A\frac{\partial u}{\partial x}\right) = 0.$$

[It is of interest to note

(i) that these equations are hyperbolic and hence that waves can propagate along the fibre,

(ii) how difficult it is to derive these equations systematically from the Navier-Stokes equations.]

*17 Viscous fluid flows steadily in the x-direction between infinite plates $y = 0, h$. Assuming two-dimensional flow in which all the variables are independent of x, show that the temperature T satisfies

$$\frac{d}{dy}\left(k\frac{dT}{dy}\right) + \mu\left(\frac{du}{dy}\right)^2 = 0.$$

Deduce that no such "fully-developed" flow can exist when k is constant and the walls are thermally insulated. If μ is constant, find T when the upper plate is driven with a constant force so that

$$\mu\frac{\partial u}{\partial y}|_{y=h} = \tau$$

and both walls are maintained at temperature T_0. If the viscosity

[7] In fact this stress is $3\mu\frac{\partial u}{\partial x}$, and 3μ is called the Trouton viscosity.

of the fluid is given by $\mu = \mu_0 e^{-\lambda T}$, where λ is a positive constant, show that

$$\frac{d^2 T}{dy^2} + \frac{\tau^2}{k\mu_0} e^{\lambda T} = 0.$$

Deduce that, even when the walls are isothermal, the solution will now only exist if T_0 is less than some critical value. (The nonexistence for large enough T_0 is called *thermal runaway.*)

*18 Show that the stagnation point flow $\mathbf{u} = (x, -y, 0)$ satisfies both the Euler and the Navier-Stokes equations. Under what circumstances is a solution of the Euler equations also a solution of the Navier-Stokes equations?

*19 Viscous fluid occupies the region above a plane rigid boundary $z = 0$ which is rotating with angular velocity Ω. Verify that there is a "similarity" solution to the Navier-Stokes equations of the form

$$u_r = \Omega r f(\xi),\ u_\theta = \Omega r g(\xi),\ u_z = (\nu\Omega)^{\frac{1}{2}} h(\xi)$$

where $\xi = z\Omega^{1/2}/\nu^{1/2}$ and f, g, h satisfy

$$f'' = f^2 + hf' - g^2,\ g'' = 2fg + hg',\ h' + 2f = 0.$$

Show that appropriate boundary conditions are

$$f(0) = 0,\ g(0) = 1,\ h(0) = 0 \text{ and } f, g \to 0 \text{ as } \xi \to \infty.$$

Remark

As we have seen in these exercises, many of the exact solutions that satisfy physically relevant boundary conditions that are known for the Navier-Stokes equations are *unidirectional.* However we will encounter an important exception to this (known as Jeffery-Hamel flow) in exercise 8 in the next chapter.

2

Boundary layers

2.1 Asymptotic methods for boundary layers

Many books on analysis and applied mathematics discuss methods for describing the way in which the solutions of a differential equation change when the coefficient of the highest derivative becomes smaller and smaller. One phenomenon that is quite common for ordinary differential equations is that the solutions can either vary or oscillate more and more rapidly as the limit is approached. Oscillating solutions may be periodic, quasi-periodic or, for nonlinear equations, chaotic. However the rapid variations may be confined to a small part of the range of the independent variable in which case we have either an *interior layer* or, more usually, a *boundary layer*. We now develop a method which can be used to find an approximate solution to such problems. Although we will not explicitly mention asymptotic expansions in this chapter, we hope that they underlie all the approximations we shall be making when we use the symbol $\simeq$. Such 'perturbation schemes' will be discussed again in Chapter 3 and the relevance of the formal theory of asymptotic expansions is discussed briefly in Appendix A (see Hinch for a more complete introduction).

First we will treat two linear equations that have explicit solutions and then we will show how the same method can be used on the Navier-Stokes equations when $Re \gg 1$.

2.1.1 An ordinary differential equation

The idea of a boundary layer can be understood by considering the solution to the simple ordinary differential equation

$$\varepsilon\frac{du}{dx} + u = x, \quad u(0) = 1. \tag{2.1}$$

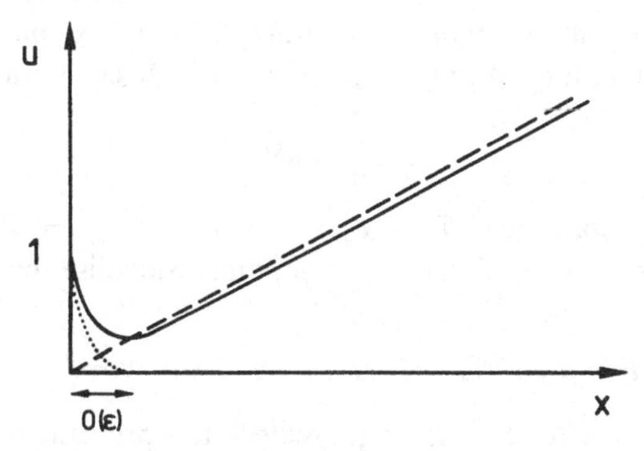

Fig. 2.1. Solution of (2.1); ... : inner solution, - - - : outer solution, —— : exact solution

This has the solution $u = (1+\varepsilon)e^{-x/\varepsilon} + x - \varepsilon$ and it is immediately clear that, if $0 < \varepsilon \ll 1$, the $e^{-x/\varepsilon}$ term is only significant when $x = O(\varepsilon)$, and that, when $x = O(1)$, we may regard u as being x to a good approximation. This can also be seen directly from (2.1) since neglecting terms of $O(\varepsilon)$ would naturally lead to the equation $u = x$. However, this cannot be an accurate solution everywhere because it does not satisfy the boundary condition $u(0) = 1$. To deal with this problem systematically we change to the *scaled variable* $X = x/\varepsilon$ so that $X = O(1)$ in a boundary layer where $x = O(\varepsilon)$ and the equation becomes

$$\frac{du}{dX} + u = O(\varepsilon), \quad u(0) = 1,$$

with solution $u = e^{-X}$ when terms of $O(\varepsilon)$ are neglected. Thus, by solving the approximate equation in the two 'regions', the *outer* region where $x = O(1)$ and the *inner* region where $x = O(\varepsilon)$, we have obtained solutions which, as shown in figure 2.1, give a good approximation to the exact solution. In particular we are able to tell *from the equation, without actually solving it*, that the derivative term is only important if $x = O(\varepsilon)$ and therefore that there may be a boundary layer of width $O(\varepsilon)$ near any point at which a boundary condition is imposed.

The linear heat conduction equation

As our next illustration, we consider the problem of heat conduction and convection in an inviscid flow. We suppose that a uniform stream with velocity $U\mathbf{i}$ and temperature T_0 flows past a 'hot' plate at $y = 0$, $x > 0$.

The plate is held at constant temperature T_1 and k is constant[1]. From the energy equation (1.24), the temperature T will satisfy the equation

$$\rho c U \frac{\partial T}{\partial x} = k\nabla^2 T, \tag{2.2}$$

with boundary conditions $T = T_0$ at infinity and $T = T_1$ on $y = 0$, $x > 0$. The first step, as always, is to nondimensionalise the variables by writing

$$T = T_0 + (T_1 - T_0)T', \quad x = Lx', \quad y = Ly'.$$

Since there is no characteristic length scale in this problem, we can choose the length L arbitrarily at this stage. Once chosen it will indicate that we are considering the flow at a distance $O(L)$ from the leading edge of the plate at $x = 0$, but it turns out that L will disappear when we express the solution in dimensional terms. The nondimensional equation is now, on dropping dashes,

$$\frac{\partial T}{\partial x} = \varepsilon\left(\frac{\partial^2 T}{\partial x^2} + \frac{\partial^2 T}{\partial y^2}\right), \tag{2.3}$$

with $T = 0$ at infinity (except near the plate when y is small and x large and positive) and $T = 1$ on $y = 0$, $x > 0$. Here $\varepsilon = \frac{1}{Pe}$ where $Pe = \frac{\rho c L U}{k}$ is the *Peclet number* which plays the same role in this conduction problem that the Reynolds number plays in the Navier-Stokes equations. We wish now to consider solutions of (2.3) when ε is small. Luckily[2] this problem has an exact solution that can be written as

$$T = \operatorname{erfc}\eta \tag{2.4}$$

where η is defined by the transformation $(\xi + i\eta)^2 = \frac{1}{\varepsilon}(x + iy)$ and $\operatorname{erfc}\alpha = \frac{2}{\sqrt{\pi}}\int_\alpha^\infty e^{-z^2}dz$ (exercise 3). Our ultimate aim here is to retrieve a suitable approximation to (2.4) by a method that can then be used on the more intractable Navier-Stokes equations. We first examine the solution (2.4) in more detail.

The region of interest, shaded in figure 2.2, occurs when $\eta \leq O(1)$ where the hot plate affects the temperature field significantly; this can be seen, but only with hindsight, from (2.4). We also want to keep x of $O(1)$ and so, since $x = \varepsilon(\xi^2 - \eta^2)$, these two conditions imply $\xi = O(\varepsilon^{-1/2})$.

[1] Note that for the purposes of this example we neglect the viscosity in the fluid entirely which certainly means the fluid could not be air or water since, for both these fluids, viscosity and heat conduction act on roughly the same length scales. It could be a physically realistic situation only if the thermal boundary layer, as developed in this section, is much wider than the viscous boundary layer that is discussed in Section 2.2.

[2] This "luck" is in fact a consequence of group theory: see Appendix B.

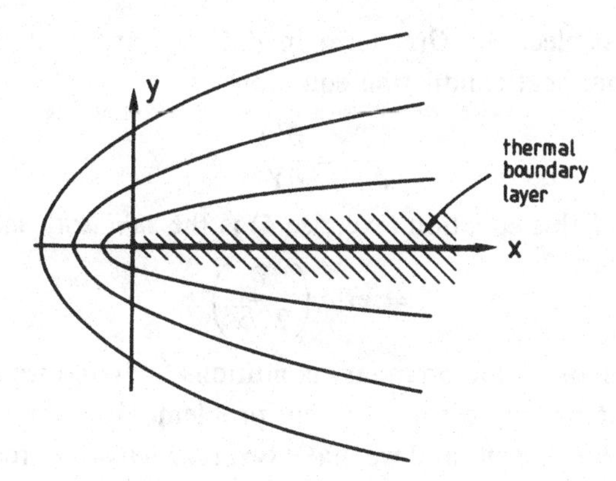

Fig. 2.2. The parabolic isotherms of the solution (2.4)

When ε is small we can therefore simplify the transformation in this region to

$$x = \varepsilon\xi^2 + O(\varepsilon), \quad y = 2\varepsilon\xi\eta$$

and hence η is approximately $\frac{y}{2\sqrt{\varepsilon x}}$ and, as long as x is not small, we can write

$$T \simeq \operatorname{erfc}\left(\frac{y}{2\sqrt{\varepsilon x}}\right). \tag{2.5}$$

This formula shows that there is a *thermal boundary layer* on the plate where $y = O(\sqrt{\varepsilon}\, x)$ in which T adjusts from the value 1 on the plate to the value zero in the external flow: also, when written in terms of our original dimensional variables, T is independent of L.

Following the ideas of the last section we now show how to obtain the scalings and the approximate solution (2.5) by scaling the variables appropriately in equation (2.3) but *without solving the full equation explicitly*. We start by writing $y = \sqrt{\varepsilon}Y$ so that (2.3) becomes

$$\frac{\partial T}{\partial x} = \frac{\partial^2 T}{\partial Y^2} + \varepsilon \frac{\partial^2 T}{\partial x^2}. \tag{2.6}$$

The boundary conditions are

$$T = 1 \text{ on } Y = 0,\ x > 0 \quad \text{and} \quad T \to 0 \text{ as } Y \to \infty \quad \text{with } x \text{ not too large,} \tag{2.7}$$

and we will need to think hard about these later.

If we now neglect the $O(\varepsilon)$ term in (2.6) we arrive at the standard one-dimensional heat conduction equation

$$\frac{\partial T}{\partial x} = \frac{\partial^2 T}{\partial Y^2}. \tag{2.8}$$

One solution of this equation (exercise 4) is the *similarity solution*

$$T = \operatorname{erfc}\left(\frac{Y}{2\sqrt{x}}\right).$$

This solution satisfies the boundary conditions (2.7) and we have arrived at (2.5) *without* having to solve the full problem. However there is more to this than meets the eye and we make two *very important* remarks about this procedure.

1. **The validity of the approximation.** When x is small we can remove ε from (2.3) by writing $x = \varepsilon X'$ and $y = \varepsilon Y'$ so that all the terms in the equation are equally important. Thus the approximate solution (2.5) is not valid *near* the leading edge of the plate though we can see by comparison with (2.4) that it is a good approximation everywhere else in the vicinity of the plate. This idea occurs again in the mechanical boundary layers considered in Section 2.2.

2. **The matching conditions.** We have taken the condition $T \to 0$ as $Y \to \infty$ from the unapproximated condition $T \to 0$ at infinity without worrying whether it really can be applied directly to the boundary layer. It is not obvious that this is the correct procedure since equation (2.8) is only valid when y is $O(\sqrt{\varepsilon})$ and, even though, $\sqrt{\varepsilon}Y = y$, the limits $y \to \infty$ and $Y \to \infty$ may not be related. In this case, because the temperature outside the boundary layer is $T \equiv 0$ to lowest order, we are able to transfer the boundary condition from $y \to \infty$ to $Y \to \infty$. However, if the temperature of the incoming fluid were *not* constant, as in exercise 5, we would need to apply a *matching condition* to connect the 'boundary layer solution' valid when $y = O(\sqrt{\varepsilon})$ with the 'outer solution' valid when $y = O(1)$.

To make this precise we need to invoke the *theory of matched asymptotic expansions* [Van Dyke (1)]. The crucial idea is that we are constructing two different asymptotic expansions for the same function $T(x, y, \varepsilon)$; the first expansion is valid away from the plate in the 'outer' region where $y \gg \sqrt{\varepsilon}$ and the second is valid in the boundary layer or 'inner' region where y is $O(\sqrt{\varepsilon})$. We expect the two solutions to join smoothly in the sense that they must both be valid and equal, to the first approximation, in some 'overlap' region where y is very small and simultaneously Y is

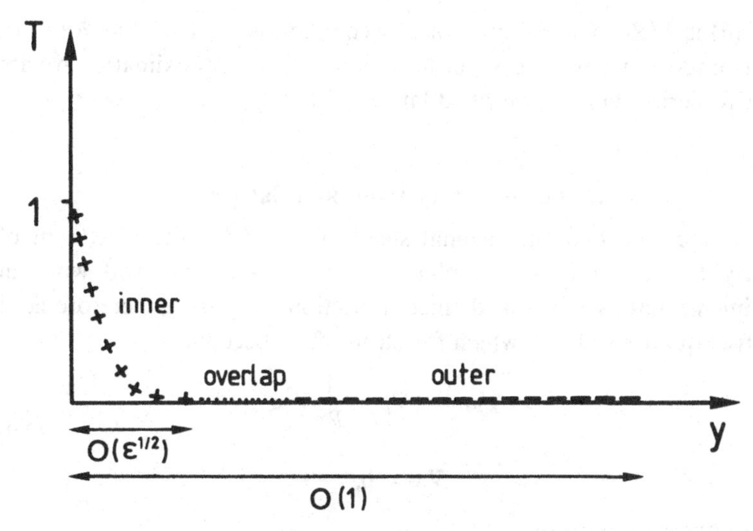

Fig. 2.3. Solution of (2.3)

very large. This will happen when $\sqrt{\varepsilon} \ll y \ll 1$; we could, for example, consider $y = O(\varepsilon^{1/4})$ as the overlap region.

If we refer back to the simple example at the beginning of this chapter we found that the outer solution to (2.1) was $u_o = x$ and the inner solution was $u_i = e^{-X}$ where $X = x/\varepsilon$.[3] If we now consider an overlap region where $\varepsilon \ll x \ll 1$ we see that both u_o and u_i become small and thus 'match' to each other.

Now, considering the present situation, T in the 'outer' region will be approximately zero in an overlap region where $y \ll 1$. The *matching condition* for the 'inner' solution where Y is $O(1)$ is therefore $T \to 0$ as $Y \to \infty$. This is illustrated in figure 2.3 where the inner solution is represented by + when $y = O(\sqrt{\varepsilon})$ and the outer solution is shown as – when $y = O(1)$.

Before proceeding to apply the idea of a boundary layer to the general Navier-Stokes equations we make one other observation. Unfortunately almost all the known exact solutions of these equations have trivial limits as $Re \to \infty$ and cannot be used as a check on our approximate methods as the exact solution (2.4) was used for the heat conduction equation. One exception is the Jeffery-Hamel solution for flow in a wedge (exercise 8) which, on putting the stream function $\psi = f(\theta)$, leads to a fourth-order ordinary differential equation for f in which the coefficient

[3] The use of suffices i and o to denote inner and outer is very useful and common.

of $f''''(\theta)$ is $1/Re$. The solution of this equation is nontrivial as $Re \to \infty$ and its importance for checking the validity of the approximation we are about to derive will be described later.

2.2 The boundary layer on a flat plate

We consider the two-dimensional steady flow of a uniform stream of velocity $U\mathbf{i}$ past a fixed flat plate $y = 0$, $x > 0$. We will work in nondimensional variables as defined in Section 1.5 and we therefore need to solve equations (1.26), which for steady flow become

$$(\mathbf{u}.\nabla)\mathbf{u} = -\nabla p + \frac{1}{Re}\nabla^2\mathbf{u}$$

$$\nabla.\mathbf{u} = 0, \tag{2.9}$$

with boundary conditions

$$\mathbf{u} = (1, 0) \quad \text{at infinity}$$

and

$$\mathbf{u} = \mathbf{0} \quad \text{on} \quad y = 0,\ x > 0.$$

Here $Re = \frac{UL}{\nu}$ where, as in the conduction problem, L is an arbitrary length scale.

In order to relate more easily to our discussion of the scalar equation (2.2), we will work with the stream function ψ, defined by $\mathbf{u} = \left(\frac{\partial\psi}{\partial y}, \frac{-\partial\psi}{\partial x}\right)$, and take the curl of (2.9) to eliminate p. The resulting equation for ψ (see exercise 7) is

$$\frac{\partial(\psi, \nabla^2\psi)}{\partial(y, x)} = \frac{1}{Re}\nabla^4\psi, \tag{2.10}$$

with boundary conditions

$$\psi \to y \quad \text{at infinity}$$

and

$$\psi = \frac{\partial\psi}{\partial y} = 0 \quad \text{on} \quad y = 0,\ x > 0.$$

When $Re = \infty$, we get the inviscid solution $\psi = y$ or $\mathbf{u} = (1, 0)$. This solution does not satisfy the no-slip condition, $\frac{\partial\psi}{\partial y} = 0$, on the plate but the inclusion of the viscous terms enables us to satisfy this condition. This is very similar to what happened in the heat conduction problem in Section 2.1; there, letting the Peclet number tend to infinity (or $\varepsilon \to 0$), the only solution was $T = 0$ which did not satisfy the condition on $y = 0$. Guided by this earlier problem, we therefore look for a *boundary layer* by scaling y with $\frac{1}{\sqrt{Re}}$ and writing

$$y = \frac{1}{\sqrt{Re}} Y. \tag{2.11}$$

In this case, since u is $O(1)$ and $\partial\psi/\partial y = u$ we will also have to scale ψ by putting

$$\psi = \frac{1}{\sqrt{Re}} \Psi. \tag{2.12}$$

Substituting for ψ, y into equation (2.10) leads to

$$\frac{\partial \Psi}{\partial Y}\frac{\partial}{\partial x}\left(\frac{\partial^2 \Psi}{\partial Y^2}\right) - \frac{\partial \Psi}{\partial x}\frac{\partial}{\partial Y}\left(\frac{\partial^2 \Psi}{\partial Y^2}\right) + \frac{1}{Re}\left[\frac{\partial \Psi}{\partial Y}\frac{\partial^3 \Psi}{\partial x^3} - \frac{\partial \Psi}{\partial x}\frac{\partial^3 \Psi}{\partial x^2 \partial Y}\right]$$
$$= \frac{\partial^4 \Psi}{\partial Y^4} + \frac{2}{Re}\frac{\partial^4 \Psi}{\partial x^2 \partial Y^2} + \frac{1}{Re^2}\frac{\partial^4 \Psi}{\partial x^4}. \tag{2.13}$$

If we now formally let $Re \to \infty$, we are left with the equation

$$\frac{\partial \Psi}{\partial Y}\frac{\partial^3 \Psi}{\partial x \partial Y^2} - \frac{\partial \Psi}{\partial x}\frac{\partial^3 \Psi}{\partial Y^3} = \frac{\partial^4 \Psi}{\partial Y^4}, \tag{2.14}$$

which holds in the boundary layer on the plate. This is a version of *Prandtl's Boundary Layer equation*[4] and we now need to solve it subject to appropriate boundary conditions. The no-slip condition on the plate is

$$\Psi = \frac{\partial \Psi}{\partial Y} = 0 \quad \text{on} \quad Y = 0, \ x > 0 \tag{2.15}$$

and we need a matching condition to apply when $Y \to \infty$. Since the outer solution in terms of y is $\psi = y$, which can be written as $\Psi = Y$, we postulate the matching condition as

$$\Psi \to Y \quad \text{or} \quad \frac{\partial \Psi}{\partial Y} \to 1 \quad \text{as} \quad Y \to \infty. \tag{2.16}$$

Before embarking on a solution of (2.14) subject to (2.15) and (2.16) we note the following three points.

[4] Prandtl's Boundary Layer equations may also be written in terms of u, v, p (see exercise 9) as

$$u\frac{\partial u}{\partial x} + V\frac{\partial u}{\partial Y} = -\frac{dp}{dx} + \frac{\partial^2 u}{\partial Y^2}$$

and

$$\frac{\partial u}{\partial x} + \frac{\partial V}{\partial Y} = 0$$

where $v = \frac{1}{\sqrt{Re}} V$.

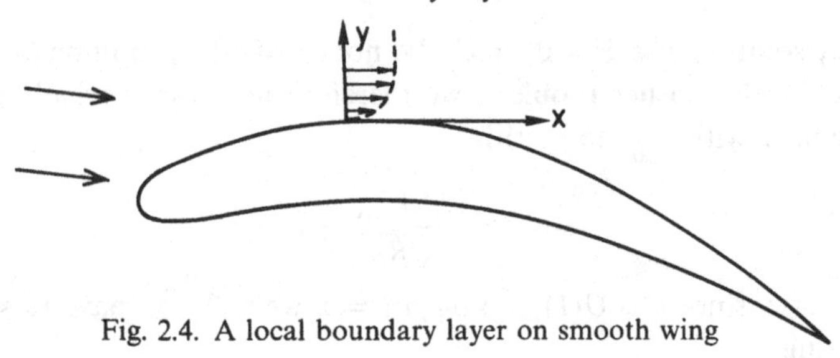

Fig. 2.4. A local boundary layer on smooth wing

1. **The pressure.** With the scalings (2.11,12), the momentum equation in the y-direction in (2.9) gives $\partial p/\partial Y = O(1/Re)$ and hence the pressure does not vary across the boundary layer to lowest order.

2. **The leading edge.** We expect to see the same 'nonuniformity' near the leading edge, $x = 0$, of the plate as occurred for the heated plate problem. If we scale x, y and ψ all with $\frac{1}{Re}$ we are left with the full problem which no one has ever solved analytically.

3. **The boundary layer on a general body.** We might expect equations very similar to (2.14) to apply in the boundary layer on an *arbitrary* smooth two-dimensional body. Away from the body the flow will be basically inviscid and hence will "slip" past the body with a velocity $U_s(x)$. If this was indeed the case then *locally* the situation in the boundary layer would be just the same as for a flat plate (figure 2.4). To apply the no-slip condition on the body, where $\psi = 0$, we would need to introduce the same scalings

$$y = \frac{1}{\sqrt{Re}} Y \quad \text{and} \quad \psi = \frac{1}{\sqrt{Re}} \Psi, \tag{2.17}$$

where y is now a coordinate[5] measured normal to the boundary. The only way that this boundary layer would 'know' that it is on a curved boundary is through the matching condition which becomes

$$\Psi_Y \to U_s(x) \quad \text{as} \quad Y \to \infty.$$

All this is only possible if the boundary is smooth and the radius of curvature sufficiently large and the flow is two-dimensional (exercise 18). However, even if these conditions *are* met, it is still possible that the boundary layer may 'separate' from the body and then, as is illustrated

[5] The coordinate x is taken as arc-length along the smooth two-dimensional body and y as an orthogonal coordinate as shown in Figure 2.4. Such a coordinate system will not necessarily be defined globally, but we will only need it in the vicinity of the body.

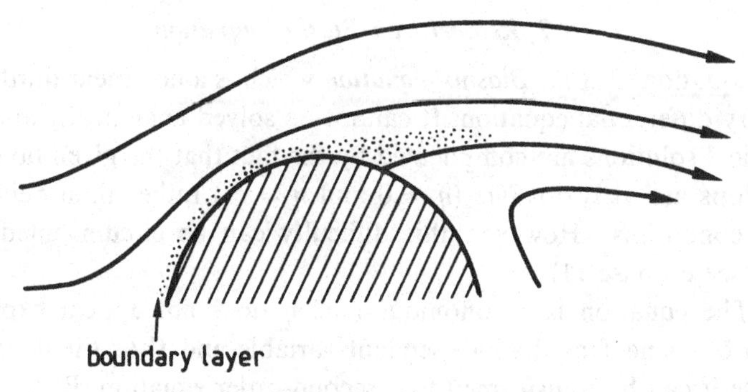

Fig. 2.5. Separation in viscous flow

in figure 2.5, the flow may differ over a large region from the inviscid limit. We shall return to this phenomenon later.

2.2.1 The solution of Prandtl's boundary layer equations for flow past a flat plate

We first observe that we can integrate equation (2.14) with respect to Y to get

$$\frac{\partial \Psi}{\partial Y}\frac{\partial^2 \Psi}{\partial x \partial Y} - \frac{\partial \Psi}{\partial x}\frac{\partial^2 \Psi}{\partial Y^2} = \frac{\partial^3 \Psi}{\partial Y^3} + g(x). \tag{2.18}$$

The boundary condition at infinity will determine $g(x)$ and, when this is uniform flow, (2.16) shows that $g(x) = 0$. Indeed, it is easy to see that for more general flows $g(x) = -\frac{dp}{dx}$. We now observe that equation (2.18) and boundary conditions (2.15), (2.16) are invariant under the transformation

$$x = \alpha^2 x', \quad Y = \alpha Y', \quad \Psi = \alpha \Psi'$$

for any parameter α. This implies that there is a similarity solution (Appendix B) to this problem and it can be turned from a partial differential equation to an ordinary differential equation by a transformation of the form

$$\Psi = x^{1/2} f\left(\frac{Y}{x^{1/2}}\right). \tag{2.19}$$

We see by substituting this expression into (2.18) that $f(\eta)$ must satisfy the ordinary differential equation

$$f''' + \frac{1}{2} f f'' = 0 \tag{2.20}$$

with $f(0) = f'(0) = 0$ and $f' \to 1$ as $\eta = \frac{Y}{x^{1/2}} \to \infty$.

2.2.2 Remarks on Blasius' equation

1. Equation (2.20) is *Blasius' equation* which is a nonlinear third-order ordinary differential equation. It cannot be solved analytically and even numerical solutions are complicated by the fact that the given boundary conditions are at *two points* ($\eta = 0$ and $\eta = \infty$) rather than being just initial conditions. However, this difficulty can be circumvented quite easily (see exercise 11).

2. The equation is autonomous (i.e., η does not appear explicitly) and so by using f as the independent variable and f' as the dependent variable it can be transformed to a second-order equation. Remarkably, a further reduction is possible for this equation leaving us eventually with a first-order equation (exercise 12).

2.2.3 Boundary layer equations in a nonuniform external flow

It is straightforward to generalize (2.18) to take account of a variable external velocity. If $\frac{\partial \Psi}{\partial Y} \to U_s(x)$ as $Y \to \infty$, $g(x) = U_s(x)U_s'(x)$ and a similarity solution can still be found if $U_s(x) = x^m$ or e^{ax} (exercise 13). When $U_s = x^m$, the ordinary differential equation obtained in place of (2.20) is

$$f''' + \frac{m+1}{2} f f'' = m(f'^2 - 1), \tag{2.21}$$

which is again an autonomous, nonlinear ordinary differential equation. The solutions of (2.21) subject to the same boundary conditions as (2.20) are tabulated in Rosenhead [p.250], and the form of the solutions is shown in figure 2.6.

Note that $u = \frac{\partial \Psi}{\partial Y} = x^m f'(\eta)$ so the f' plots have immediate physical significance as tangential velocity as a function of normal distance. It can be shown that when $m > 0$ there is a unique solution and for $m < -0.0904$ there is no solution. When $-0.0904 < m < 0$, however, two solutions are possible, one of which exhibits a region of reverse flow where $f'(\eta)$ is negative. This latter situation is contrary to physical expectations and is related to the fact that, if $m < 0$, the external flow is decelerating along the body. Indeed, we also note that the pressure in the boundary layer is a function only of x and can be evaluated from the outer inviscid flow equations, where Bernoulli's equation gives $\frac{dp}{dx} = -g(x) = -U_s U_s'$. Thus the sign of U_s' will determine whether there is a *favourable* ($U_s' > 0$) or *adverse* ($U_s' < 0$) pressure gradient and the latter case will occur when $U_s = x^m$ and $m < 0$. For example,

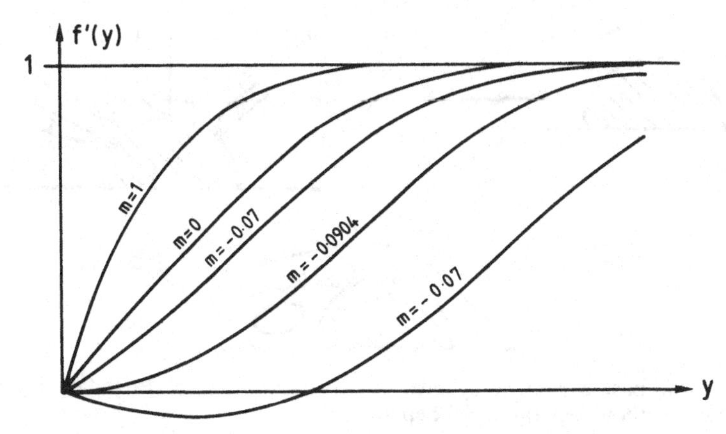

Fig. 2.6. Solutions of (2.21)

if we consider flow past a circular cylinder it can be shown (exercise 10) that if x is measured along the circumference of the cylinder, the external inviscid flow predicts a dimensionless slip velocity $U_s(x) = 2\sin x$. Thus the velocity increases on the upstream surface of the cylinder but then decreases on the downstream surface. The numerical solution of the boundary layer equations is as shown in figure 2.7b and flow reversal (where $\mu\frac{\partial u}{\partial y}|_{y=0} = 0$) occurs at $x = 1.815..$ which is only a short distance downstream of the point where the adverse pressure gradient sets in.

2.2.4 *Limitations of Prandtl's equation*

The above considerations lead us to a prediction which is borne out by observation: the physically unrealistic reverse flow profile of figure 2.7b suggests that boundary layer theory becomes inapplicable downstream of $x = 1.815$ and in practice "separation" is observed in such a configuration as indicated in figure 2.7c.[6]

It is only relatively recently that the details of the so-called 'triple deck' flow near such a separation point have been elucidated. We will not describe the full solution here but we will outline the structure of the solution and the scalings required. This involves the ideas of matched

[6] A dramatic result of the theory of boundary layer separation is the observation that while it is possible to *blow* out a candle, it is almost impossible to *suck* one out. This is because separation from the lips causes a jet to form between more or less confined shear layers and this jet can hence be aimed at the target. Suction on the other hand is inevitably omnidirectional and hence it has been claimed that there is no such thing as a point source in inviscid fluid dynamics! (See §5.3.)

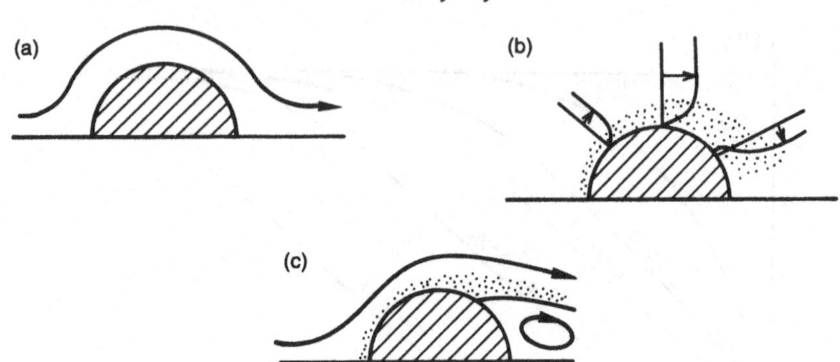

Fig. 2.7. Flow past a circular cylinder (a) Inviscid Flow (b) Boundary layer development without separation (c) Separation

asymptotic expansions that have already been applied in deriving the boundary layer but things are *much more* complicated here! Indeed, the next two pages are inessential for the rest of the notes and should only be studied by devotees of boundary layer theory.

The basic situation we consider is when a flow with a Blasius type of profile (figure 2.6) approaches a point $x = 0$ where the flow is about to change direction. The key idea is that the fluid will have reverse flow for $x > 0$ in a layer that is very close to the wall and, although thinner than $Re^{-1/2}$, it will still be described by a Prandtl boundary layer equation with no slip at the wall. Above it there will be another boundary layer also satisfying the boundary layer equations but being pushed away from the wall by the reverse flow layer. Suppose that the 'lower deck' layer has thickness $Y = O(\delta)$ and is of streamwise extent $x = O(l)$ where δ and l are both small quantities. Going upstream from this layer u is proportional to Y in the Blasius boundary layer so $u = O(\delta)$ and Ψ will be $O(\delta^2)$. Thus, if Prandtl's boundary layer equations are to hold, the terms in (2.14) must still balance and so $l = O(\delta^3)$. From the equation in the footnote on p. 29 we can also estimate that $p = O(\delta^2)$ in this layer. Thus equation (2.14) will hold in the lower deck with the no slip condition at the wall and a matching condition with the solution in the layer above as $Y \to \infty$. This second 'main deck' Prandtl boundary layer is of the usual width $Re^{-1/2}$ and continues directly from the boundary layer in $x < 0$ but it will see both a *slip velocity* on the wall and a *displacement* perpendicular to the wall in $x > 0$. This structure is sketched in figure 2.8 and two key results follow:

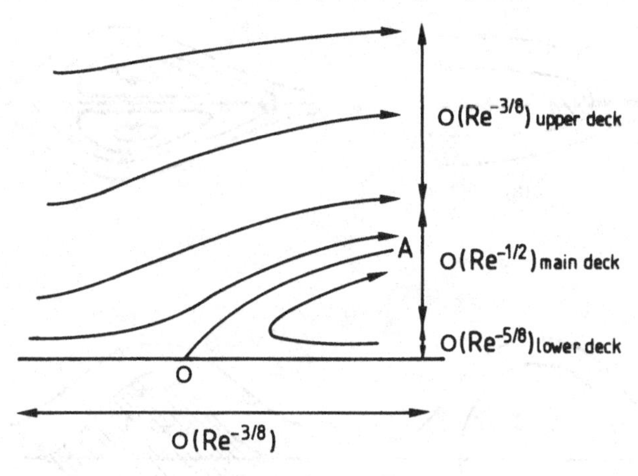

Fig. 2.8. Flow near separation

1. The 'main deck' layer makes a small angle $O\left(\frac{\delta\, Re^{-1/2}}{l}\right) = O\left(\delta^{-2} Re^{-1/2}\right)$ with the free stream[7] and hence the almost inviscid outer flow (or 'upper deck') must adjust its pressure by the same amount in response. This follows from the fact that the linearized theory of inviscid flow past an aerofoil of thickness t predicts a pressure variation of $O(t)$. But we have already noted that $p = O(\delta^2)$ in the lower deck and so equating these two pressure estimates across the main deck (where p only depends on x) gives $\delta = Re^{-1/8}$. Hence the width of the lower deck is $y = O(Re^{-5/8})$.

2. On the separating streamline OA, since $Y = O(\delta)$ and $x = O(l)$, $Y = O(x^{1/3})$ and the slope of this line will tend to zero downstream. The triple deck structure occurs in a very small region as far as the outer flow is concerned and the above observation seems to imply that the only large scale inviscid outer flow to which the 'main deck' can match is the so-called *Helmholtz flow* in which the separation stream line leaves the body tangentially and separates outer inviscid flow from an essentially constant pressure region or 'cavity' as shown in figure 2.9a.

We refer those interested in the full theory of 'triple decks' to Smith. A mathematical complication is caused by the fact that the outer transverse velocity and pressure seen by the 'main deck' are related by a "Hilbert transform" integral solution of the inviscid upper deck (exercise 15). This

[7] The factor δ/l comes from the lower deck and $Re^{-1/2}$ from the basic boundary layer scaling.

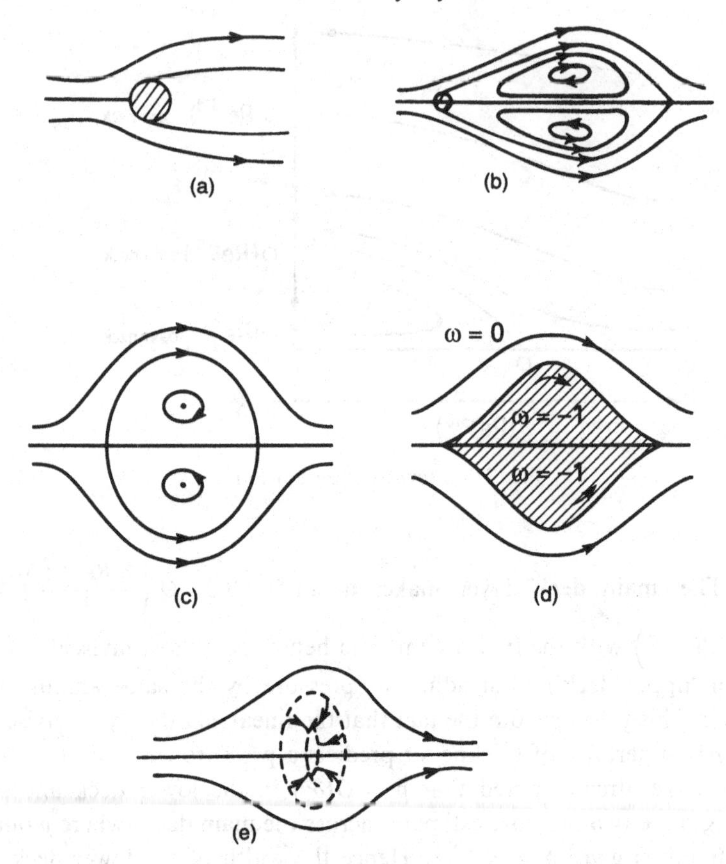

Fig. 2.9. Theories for global separation

integral relation coupled with the main deck Prandtl problem constitutes a formidable mathematical challenge.

Recent large scale numerical calculations [Fornberg] have shed quite a new light on the general problem of steady[8] bluff body flows at high *Re*. Whereas the Kirchhoff flow incorporates a stagnant cavity that grows parabolically downstream, these calculations show quite a different kind of steady wake. Not only is the velocity in the wake non-zero, being of constant small vorticity, but the length of the wake is $O(Re)$ and has finite aspect ratio (length/width) of approximately 1.669 (figure 2.9b). Hence, if this steady flow were realised, the body should leave a large,

[8] We will see in the final chapter that it is in fact unrealistic even to pose the problem of such a steady high *Re* flow because of the instabilities that engender unsteadiness as *Re* is increased through certain critical values.

wide, and weak wake.[9] This scenario can be put into a wider perspective if we note that a pair of equal and opposite inviscid vortices can remain at rest in a uniform irrotational stream (figure 2.9c). Now, if we "fatten" these vortices into regions of uniform, finite vorticity (the axisymmetric generalisation would be to fatten a circular line vortex into a smoke ring as in figure 2.9e), we can eventually construct a flow in which the equal and opposite vortex patches join up to form a so-called Prandtl-Batchelor flow (figure 2.9d). When this is done, it is found that the aspect ratio is again 1.669.., suggesting that this constant vorticity flow is the "outer limit" of the flow past any blunt body at distances of $O(1/Re)$ (figure 2.9b).

It should be noted that if a large wake of any type forms, it affects the *whole of the external* flow including the boundary layer upstream of the separation point. This can be seen clearly in figure 2.7 where the basic inviscid flow assumed in figure 2.7a will be significantly different from the inviscid flow outside the boundary layer shown in figure 2.7c. This *global influence* is a consequence of the ellipticity of Laplace's equation and it also means that the separated flow influences the boundary layer upstream of the separation point.

2.2.5 *Predictions of Prandtl's theory*

A well-documented check on the predictions of Prandtl's theory can be obtained by evaluating the drag on a flat plate. In dimensional terms, the drag per unit length is

$$
\begin{aligned}
\sigma_{12} &= \mu\left(\frac{\partial u}{\partial y}+\frac{\partial v}{\partial x}\right)_{y=0} = \mu\frac{\partial u}{\partial y}\Big|_{y=0} \\
&= \mu\frac{U}{L}\sqrt{Re}\frac{\partial^2\Psi}{\partial Y^2}\Big|_{Y=0} \\
&= \mu\frac{U}{L}\sqrt{Re}\left(\frac{L}{x}\right)^{1/2} f''(0) \\
&= \rho(\nu U^3)^{1/2} f''(0) x^{-1/2}.
\end{aligned}
$$

Thus, as we hoped, the drag force is independent of the arbitrarily chosen length L. If we assume that this formula for the shear stress is valid for the whole of a plate of length l (i.e. we ignore the effect of both leading and trailing edges) the total drag on the top of the plate is

[9] The calculation reveals the same configuration for axisymmetric bodies except that the vorticity is proportional to the distance from the axis of symmetry.

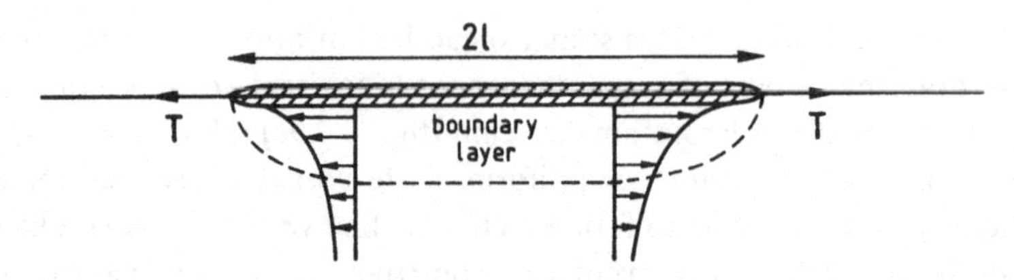

Fig. 2.10. Spreading oil film

$2\rho(\nu U^3 l)^{1/2} f''(0)$. The value of $f''(0)$ is found from numerical calculations as in exercise 11 and then this formula can be compared with experiment.

An everyday example which also validates the above theory concerns the spreading of an oil film on the surface of a pond. We consider the film as two-dimensional (figure 2.10) and assume that the spreading is due to a constant *surface tension* force T which acts on the edge of the film as shown.[10] In this case the oil film acts as two "noslip", but growing, flat plates with velocity $\pm U$ and the pond is at rest so that a boundary layer develops under the film. From the formula above we get

$$T = K U^{3/2} l^{1/2}$$

where $2l$ is the width of the film and $K = 2\rho\nu^{1/2} f''(0)$. Now $U = \frac{dl}{dt}$ and T is constant so we get

$$\frac{dl}{dt} = \left(\frac{T}{K}\right)^{2/3} l^{-1/3}$$

which leads to the expression

$$l = \left(\frac{4}{3}\right)^{3/4} \left(\frac{T}{K}\right)^{1/2} (t+A)^{3/4}, A = \text{const.},$$

for the width of the spreading film. This too can be checked against experiment.

2.2.6 The theory of flight

We are now in a position to explain a crucial aspect of the theory of flight, namely the fact that all efficient subsonic wings have sharp trailing edges as in figure 2.4. First we note that we can solve for inviscid flow outside a wedge of angle 2α by using the complex potential $w = -Uiz^{\pi/2(\pi-\alpha)}$ where $z = x + iy$ which gives $\psi = -Ur^{\pi/2(\pi-\alpha)} \cos\left(\frac{\pi\theta}{2(\pi-\alpha)}\right)$. This vanishes

[10] In fact there is an interesting dimple at the edge producing a hairline effect visually: it is known as a Thoreau wave after one of its early observers.

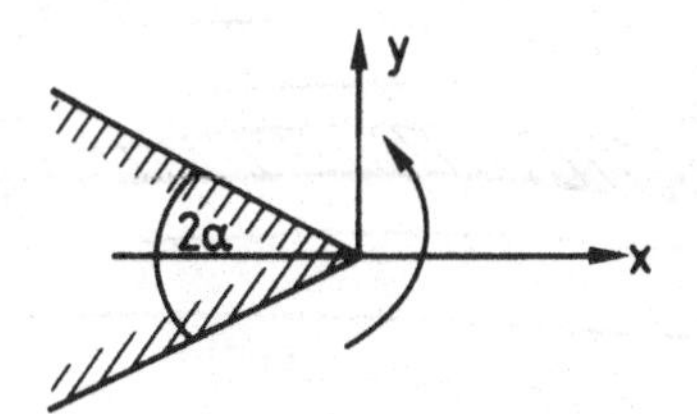

Fig. 2.11. Inviscid flow round a wedge

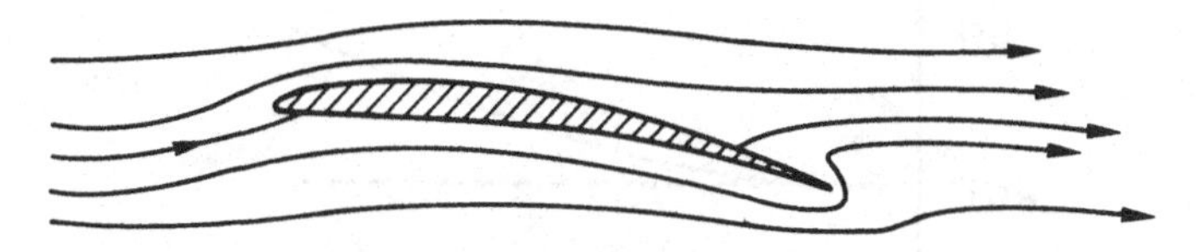

Fig. 2.12. Inviscid flow round an aerofoil

on $\theta = \pm(\pi - \alpha)$ and represents a flow round a wedge as shown in figure 2.11. On the wedge, $U_s \propto r^{-(\pi-2\alpha)/2(\pi-\alpha)}$, where r is the distance from the apex of the wedge. Now as the flow approaches the apex of the wedge along the lower surface, r is decreasing and so U_s is increasing whereas on the upper surface, where the flow is away from the apex, U_s decreases and the boundary layer is very likely to separate by the arguments adduced above.

If we now consider a two-dimensional aerofoil at incidence in a steady stream, a possible inviscid solution is as shown in figure 2.12 where we have assumed there is no circulation around the aerofoil. There will then be no lift or drag exerted on the wing (the D'Alembert Paradox) and the only way to make the inviscid model compatible with a lift on the wing is to introduce a circulation Γ at infinity. Inviscid theory shows that this will produce a lift $\rho\Gamma U$ but the solution is not unique since Γ appears to be arbitrary. However, we have just shown that the flow around the sharp trailing edge of the wing in figure 2.12 is unrealistic[11] since such a flow would separate from the upper surface, very close to the trailing edge. A more realistic picture is therefore the one shown in figure 2.13. Thus it is the *viscous* effects in the boundary layer that provide the theoretical basis for the *Kutta-Joukowski* hypothesis that the circulation in *inviscid* flow is such as to ensure smooth separation of the flow at the trailing edge.

This theory provides a very nice example of nonuniform convergence. If we plot the lift of a steady two-dimensional wing as a function of

[11] But we will see in Chapter 4 a situation where this flow is realistic!

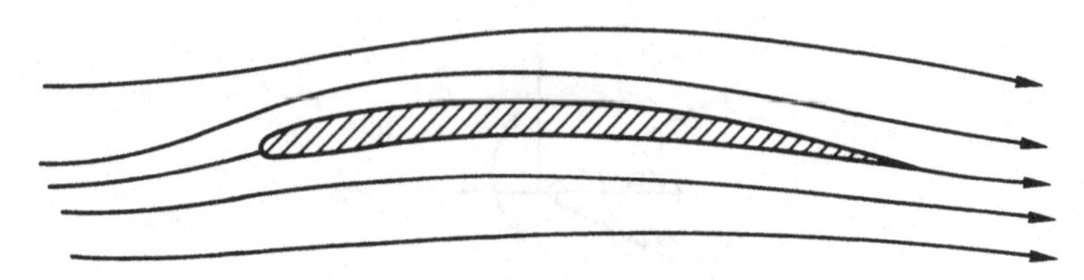

Fig. 2.13. Realistic flow around an aerofoil

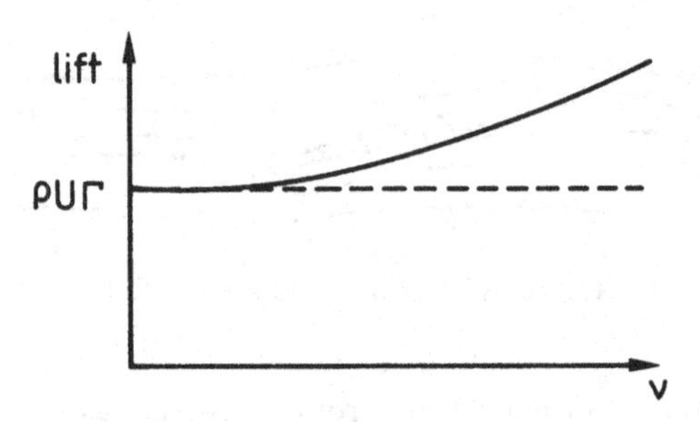

Fig. 2.14. Lift as a function of v

v in figure 2.14 we see that although the lift is zero when $v = 0$ (by D'Alembert) it is $\rho U\Gamma$ as $v \downarrow 0$. By the argument above, this lift is due to viscosity but not dependent on the size of viscosity. It is a remarkable illustration of the nonuniform convergence of the solution of the Navier-Stokes equations as $Re \to \infty$ that a boundary layer of thickness 10^{-1}cm (exercise 16) is crucial for an airliner to cross the Atlantic!

2.3 Further applications of Prandtl's equations

2.3.1 Thin wakes

The wake, mentioned above, that will develop behind a body moving through a viscous fluid is not usually susceptible to even large scale computing since the flow becomes turbulent (see Chapter 5). However, for a *thin* wake behind a thin aerofoil we may expect that Prandtl's equations will still hold. As can be seen from figure 2.15, the only difference between the flow in the wake and in the boundary layer is in the boundary conditions. If we consider the wake behind a symmetric wing, then condition (2.15) of no slip on the body is replaced by a symmetry condition that

$$\Psi = 0, \quad \text{and} \quad \frac{\partial u}{\partial Y} = \frac{\partial^2 \Psi}{\partial Y^2} = 0 \quad \text{on} \quad Y = 0. \tag{2.22}$$

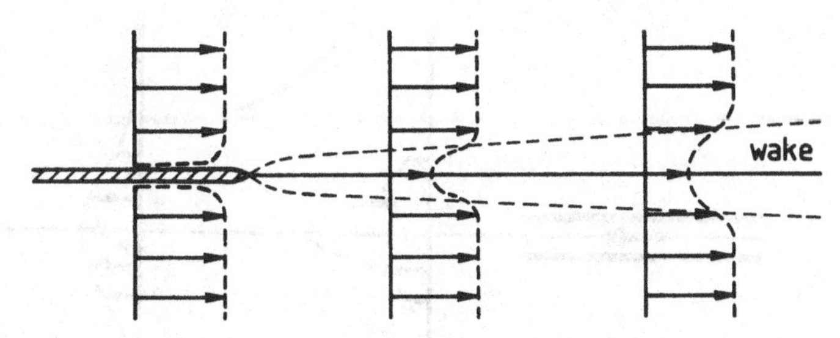

Fig. 2.15. A thin wake behind a flat plate

An explicit nontrivial solution to this problem is difficult to obtain except *far* downstream when u is close to the free stream value 1. In this case, by writing $\Psi = Y + \psi'$ and neglecting squares of ψ' and its derivatives in (2.18) with $g(x)$ put equal to zero, we are led to the linear equation

$$\frac{\partial u'}{\partial x} = \frac{\partial^2 u'}{\partial Y^2}$$

where $u' = \frac{\partial \psi'}{\partial Y}$. This is the heat equation and it can easily be seen that $u' = A\frac{1}{\sqrt{x}}e^{-Y^2/4x}$ is a suitable solution. The constant A can be determined by matching with the solution of the wake *near* the body which is a much more complicated problem. This is because rapid streamwise changes occur near the trailing edge of the wing and so Prandtl's equations, which were based on the fact that derivatives across the boundary layer dominate streamwise derivatives, are no longer valid there. Here we have a similar problem to that encountered in our discussion on separation. Indeed it was analysis of this trailing edge problem that paved the way for the triple deck theory. Again there are three different regions to be considered and the scalings needed are exactly the same as those shown in figure 2.8 although the boundary conditions are different. It is remarkable that at no stage do we need to solve the full Navier-Stokes equations. In fact it turns out that we do not need to solve this intricate problem to determine A. We can instead use a global argument to show that A is $\frac{1}{2\sqrt{\pi}}$ times the dimensionless drag on the body.

2.3.2 Jets

If a narrow jet is injected into an infinite expanse of the same fluid in such a way as to 'entrain' fluid, the situation is somewhat similar to that of a thin wake. Suppose that a directed source emits a thin two-dimensional sheet into an unbounded quiescent fluid as shown in figure 2.16. We

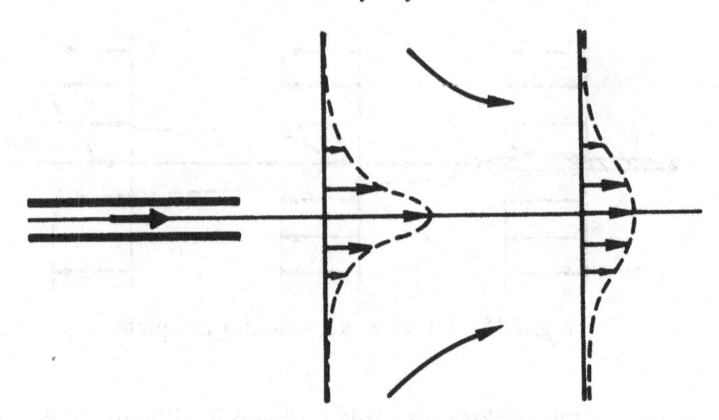

Fig. 2.16. A jet in an unbounded fluid

again expect Prandtl's equations (2.14) to apply in the jet with, in this case, the boundary conditions

$$\Psi = \frac{\partial^2 \Psi}{\partial Y^2} = 0 \quad \text{on} \quad Y = 0 \quad \text{and} \quad \frac{\partial \Psi}{\partial Y} \to 0 \quad \text{as} \quad Y \to \infty.$$

To obtain a nontrivial solution we must impose a condition which measures the strength of the jet. One possibility might be to try to prescribe the mass flow

$$\int_{-\infty}^{\infty} u \, dY$$

but because of entrainment this quantity is *not* constant. However, since the pressure is constant throughout the fluid, we can consider the x-momentum of the fluid, which is conserved in $x > 0$, by integrating equation (2.18) to obtain

$$\int_{-\infty}^{\infty} (\Psi_Y \Psi_{xY} - \Psi_x \Psi_{YY}) dY = [\Psi_{YY}]_{-\infty}^{\infty} = 0.$$

Then, integration by parts shows that

$$\int_{-\infty}^{\infty} u^2 dY = \text{constant}$$

and it is possible to find an explicit similarity solution to this problem (exercise 17).

We note that the pressure is also constant in the wake flow discussed earlier but that in this case the linearisation of the above formula implies that the mass flow $\int_{-\infty}^{\infty} u' dY$ is constant.

2.3.3 More general boundary layers

We first note that all the preceding theory has been presented for two-dimensional steady flow, for simplicity. Three-dimensional boundary layers are very important, say near wingtips and at fuselage/wing junctions, and unsteady boundary layers are crucial in understanding the dangerous phenomenon of "flutter". Throughout all these studies, the most useful information to emerge from boundary layer theory is the indication it gives about the gross features of the outer inviscid flow. For example, the initiation of wingtip[12] vortices or leading edge vortices on delta wings requires information about the separation of three-dimensional boundary layers. Also the question of whether the Kutta-Joukowski condition holds in unsteady flow is still receiving attention.

It is not surprising that the asymptotic ideas behind the theory of boundary layers developed by Prandtl for high Reynolds number flows have been found to apply throughout applied science. Here we list some other fluid dynamical applications.

1. **Thermal or diffusion boundary layers.** We have already seen an elementary example in Section 2.1 but models involving heat and mass transfer describe many familiar phenomena such as crystal growth, filters, natural and forced convection in buoyant plumes, and plate tectonics. The solution of the boundary layer equations is particularly valuable in determining the net heat or mass flux and the global evolution of the flow field (Carslaw and Jaeger).

2. **Shock waves in gases and bores in hydraulics.** In a shock wave, the viscous action is typically confined to a region which is *very* thin compared to the length scales in the background flow. Since the inviscid flow is now described by a hyperbolic system of equations it is possible to have discontinuities or 'shocks' in the inviscid flow whose interior structure can be described by including viscosity and using the ideas of boundary layer theory. The solutions of the boundary layer equations are used to relate the variables on either side of the shock and also to determine the position of the shock which is not known a priori. A similar situation is when a bore or jump occurs in fast flowing shallow water (e.g., the Severn estuary or the kitchen sink), but in these flows the "boundary layer" mechanisms may be difficult to elucidate (Lighthill).

[12] It is a common observation in the stratosphere that wingtips shed counter-rotating vortices. At ground level cars (although not good lifting bodies!) also shed vortices as can be seen from the dirty lines that develop in the middle of the carriageway.

3. **Flows with strong rotation or stratification.** The effects of rotation and stratification introduce large *undifferentiated* terms into the Navier-Stokes equations and can cause surprising effects. In particular, Ekman layers which form on the base of a rotating fluid container can be analysed by boundary layer theory. These flows are often of great environmental significance.

It is also well known that stirring a coffee cup with a spoon is much more efficacious than rotating the cup. This can be understood by noticing that the vorticity that is created at the edge of the spoon (as at a wingtip) is much greater than that created in the boundary layer at the edge of the rotating cup (Greenspan). Strong rotation about an axis parallel to the free stream can also dramatically change the structure of the wake described in Section 2.2.4; indeed, the Coriolis term in the high-Reynolds number Navier-Stokes equations can explain the existence of both downstream and *upstream* wakes in such flows (Lighthill).

4. **Flows with strong chemical reactions.** The presence of rapidly varying volumetric source terms in the energy or mass transport equations can result in an interior layer similar to a shock wave. The modelling of the chemical kinetics makes this a complicated problem but a theory can be developed to describe many real-world effects. The theory of flames, modelled as very thin reacting layers separating the burnt from the unburnt gas, is one such example (Buckmaster and Ludford).

5. **Lubrication theory.** As we shall see in Chapter 4, lubrication theory is developed for a thin layer of very viscous fluid. The methods of boundary layer theory can be applied but in this case the thickness of the layer will be prescribed by the geometry of the bearing (Langlois).

6. **Acoustics.** Viscosity is rarely relevant in the flow of small amplitude motion in gases but localized effects near corners or caustics can be important. Sometimes this demands a new scaling in a 'square' high intensity region rather than a thin layer but the scaling ideas are just the same as those used in boundary layer theory (Landau & Lifshitz).

Exercises

1 Solve the ordinary differential equation

$$\varepsilon u'' + (1-\varepsilon)u' - u + 1 = 0 \quad (\varepsilon > 0)$$

explicitly when $u(0) = 0$, $u(\infty) = 1$.

If $\varepsilon \ll 1$, use a simple perturbation scheme to find approximate solutions near $x = 0$ and away from $x = 0$. Compare the approximate and exact solutions.

2 Solve $\varepsilon u'' + u' = 2$, $u(0) = 0$, $u(1) = 1$ where ε is a small positive constant. Show that there is a boundary layer on $x = 0$ but not on $x = 1$. Could you have anticipated this without knowing the exact solution?

3 If variables ξ, η are defined by $(\xi + i\eta)^2 = \frac{1}{\varepsilon}(x+iy)$, show that it is possible to find a similarity solution of (2.3) by writing $T = f(\eta)$ and that f satisfies

$$f'' + 2\eta f' = 0$$

with $f(\infty) = 0$ and $f(0) = 1$. Hence show that $T = \frac{2}{\sqrt{\pi}} \int_\eta^\infty e^{-z^2} dz$ $= \operatorname{erfc} \eta$.

4 Show that the problem

$$\frac{\partial T}{\partial x} = \frac{\partial^2 T}{\partial Y^2}$$

with $T(x,0) = 1$ and $T(x,\infty) = 0$ has a similarity solution $T = F(\zeta)$ where $\zeta = \frac{Y}{2\sqrt{x}}$ and that its solution is $T = \operatorname{erfc}\left(\frac{Y}{2\sqrt{x}}\right)$.

5 Use boundary layer methods to solve equation (2.3) for heat conduction in a flowing medium when the temperature $T \to 2 + 3y$ as $y \to \infty$ and $T = 0$ on $y = 0$, $x > 0$.

*6 Inviscid fluid flows with constant velocity $U\mathbf{i}$ past a semi-infinite flat plate $y = 0$, $x > 0$. Explain why the temperature T satisfies

$$\frac{\partial T}{\partial t} + U\frac{\partial T}{\partial x} = k\nabla^2 T.$$

The temperature in the fluid is T_0 and the plate is suddenly heated to $T = T_1$ for $t > 0$. Show that in the 'temperature boundary layer' near the plate

$$\frac{\partial \bar{T}}{\partial t} + U\frac{\partial \bar{T}}{\partial x} = \frac{\partial^2 \bar{T}}{\partial Y^2}$$

(where $\bar{T}, Y$ are suitably scaled variables) and $\bar{T} = 1$ on $Y = 0$, $\bar{T} = 0$ at $t = 0$ and as $Y \to \infty$.

Verify that this inner problem has a similarity solution

$$T = \begin{cases} \operatorname{erfc}(Y/2\sqrt{t}) & \text{if } \; Ut < x \\ \operatorname{erfc}(Y/2\sqrt{\frac{x}{U}}) & \text{if } \; Ut > x. \end{cases}$$

How far away from the leading edge $x = 0$ must an observer on the plate be if he is to be insensitive to its presence?

7 Show that in two dimensions $\mathbf{u}$ can be written in terms of a stream function ψ and the Navier-Stokes equations (2.9) lead to the equation

$$\frac{\partial(\psi, \nabla^2\psi)}{\partial(y, x)} = \frac{1}{Re}\nabla^4\psi.$$

8 Show that an exact two-dimensional solution of (2.10) can be found by writing $\psi = f(\theta)$ (θ is the polar angle) where f satisfies

$$\frac{1}{Re}(f'''' + 4f'') + 2f'f'' = 0.$$

Write down the appropriate boundary conditions for flow in a wedge with fixed walls at $\theta = 0, \alpha$ (Jeffery-Hamel flow).

9 By scaling the variables in (2.9) appropriately show how Prandtl's boundary layer equations

$$\left.\begin{aligned} u\frac{\partial u}{\partial x} + v\frac{\partial u}{\partial y} &= -\frac{dp}{dx} + \frac{1}{Re}\frac{\partial^2 u}{\partial y^2} \\ \frac{\partial u}{\partial x} + \frac{\partial v}{\partial y} &= 0 \end{aligned}\right\} \tag{2.23}$$

can be derived when $Re \gg 1$.

10 A circular cylinder of radius a is placed in uniform stream with velocity U_∞. If $Re = \frac{Ua}{\nu} \gg 1$, show that if there is a boundary layer on the cylinder, the velocity outside this layer will be $U_s(x) = 2U_\infty \sin\frac{x}{a}$ where x is arc length on the cylinder measured from the leading generator.

11 Show that the scaling $f = \alpha F(\zeta)$ where $\eta = \frac{\zeta}{\alpha}$ leaves the Blasius equation (2.20) unchanged but changes the boundary condition at infinity. Suppose that the solution of the initial value problem

$$F''' + \frac{1}{2}FF'' = 0$$

with $F(0) = F'(0) = 0$, $F''(0) = 1$ is calculated and it is found that $F'(\infty) = C$, which is approximately $(0.332)^{-2/3}$. How can $f(\eta)$ satisfying (2.20) be found in terms of F and what is the value and physical significance of $f''(0)$?

12 Show that equation (2.21) can be transformed into a second-order equation for $u(z)$ by writing $u = f'(x)$, $z = f(x)$. To what do the boundary conditions transform?

When $m = 0$, show that the second-order equation can be made autonomous by writing $u = z^2 g(\log z)$ and hence reduce it to a first-order equation for g'. What are the boundary conditions now?

Can you suggest any reason why these "order-lowering" transformations should exist? (See Appendix B.)

13 The boundary layer equations for a variable external velocity $U_s(x)$ are

$$\frac{\partial \Psi}{\partial Y}\frac{\partial^2 \Psi}{\partial x \partial Y} - \frac{\partial \Psi}{\partial x}\frac{\partial^2 \Psi}{\partial Y^2} = \frac{\partial^3 \Psi}{\partial Y^3} + U_s(x)U_s'(x)$$

subject to $\frac{\partial \Psi}{\partial Y} = \Psi = 0$ on $Y = 0$ and $\frac{\partial \Psi}{\partial Y} \to U_s(x)$ as $Y \to \infty$. Show that a similarity solution of the form

$$\Psi = \alpha(x) f(\eta)$$

where $\eta = \frac{Y}{\beta(x)}$ and where α and β are functions to be determined, can be found only when $U_s(x) = \kappa(x + c)^m$ or $\kappa e^{a(x+c)}$ (κ, m, a, c constant). Write down the equation and boundary conditions satisfied by f in each case.

14 Starting from the equation derived in exercise 8, show that for radial flow in a wedge ($0 < \theta < \alpha$) the velocity is $(g(\theta)/r)\mathbf{e}_r$ where g satisfies

$$\frac{1}{Re}(g'' + 4g) + g^2 = \text{constant}$$

with $g(0) = g(\alpha) = 0$. If there is a source of strength Q at the vertex, show that an approximate "outer" solution when $Re \gg 1$ might be $g \sim Q/\alpha$. Show further that in any boundary layer that might exist on $\theta = 0$ the appropriate equation for g is

$$\frac{d^2 g}{d\phi^2} + g^2 = \frac{Q^2}{\alpha^2}$$

where $\phi = Re^{\frac{1}{2}}\theta$. Write down the boundary and matching conditions for g and deduce that such a solution is only possible for in-flow (i.e., $Q < 0$). By rewriting the boundary value problem in terms of x, Y show that the above solution is identical to the similarity solution found in exercise 13 with $m = -1$ and that the condition $Q < 0$ corresponds to the fact that stable flow is only expected when $U_s'(x) > 0$.

*15 The inviscid 'upper deck' of a triple deck flow is modelled by steady two-dimensional potential flow in $y > 0$ where matching with the 'main deck' gives one relation between the tangential velocity $u(x)$ (and hence the pressure) and the normal velocity $v(x)$ on $y = 0$. A second relation can be found by the following procedure:

(i) Show that

$$\phi = \frac{1}{2\pi}\int_{-\infty}^{\infty} v(\xi)\log[(x-\xi)^2 + y^2]d\xi$$

satisfies Laplace's equation in $y > 0$ and that

$$\frac{\partial\phi}{\partial y} = \frac{y}{\pi}\int_{-\infty}^{\infty}\frac{v(\xi)}{(x-\xi)^2+y^2}d\xi$$

tends to $v(x)$ as $y \downarrow 0$ (note the dominant contribution to the integral comes from near $\xi = x$ when y is small).

(ii) Show that

$$u(x) = \lim_{y\downarrow 0}\frac{\partial\phi}{\partial x} = -\frac{1}{\pi}\fint_{-\infty}^{\infty}\frac{v(\xi)d\xi}{\xi - x}$$

where the integral on the right-hand side is the *Cauchy Principal Value* defined as $\lim_{\varepsilon\to 0}\left(\int_{-\infty}^{x-\varepsilon} + \int_{x+\varepsilon}^{\infty}\right)\frac{v(\xi)d\xi}{\xi - x}$ and $-u$ is called the *Hilbert Transform* of v.

16 (i) Estimate the size of the Reynolds number for flow over the wing of a jumbo jet. (The kinematic viscosity of air at 30,000 feet is about $10^{-1}\text{cm}^2/\text{sec}$.)

(ii) Show that the complex potential $w = -iz^{\frac{1}{2}}$ $(\text{Rl}(z^{\frac{1}{2}}) \geq 0)$ describes irrotational inviscid flow around a semi-infinite flat plate $y = 0$, $x \leq 0$. Show that the pressure gradient on the plate is $\frac{dp}{dx} = -\frac{1}{8x^2}$. How would you expect the solution of Prandtl's equations in the boundary layer on the plate to differ in $y > 0$ and $y < 0$?

Comment on any implications for the theory of flight.

17 Show that a similarity solution of the equations for the jet described in §2.3 may be found in the form

$$\psi = \nu^{1/2}x^{1/3}f(\eta), \quad \eta = y/3\nu^{1/2}x^{2/3},$$

where ψ is the streamfunction. Find the differential equation satisfied by f, and hence show that $f(\eta) = 2\alpha \tanh \alpha\eta$, where $\alpha = (9M/16\nu^{1/2})^{1/3}$.

*18 Show that if ζ is a coordinate normal to a curved body given by $\mathbf{r} = \mathbf{r}(\xi, \eta)$ on which ξ, η are orthogonal curvilinear coordinates, then Prandtl's equations in a boundary layer on the body become

$$\frac{u}{h_1}\frac{\partial u}{\partial \xi} + \frac{v}{h_2}\frac{\partial u}{\partial \eta} + w\frac{\partial u}{\partial \zeta} + \frac{uv}{h_1 h_2}\frac{\partial h_1}{\partial \eta} - \frac{v^2}{h_1 h_2}\frac{\partial h_2}{\partial \xi} = \frac{-1}{h_1}\frac{\partial p}{\partial \xi} + \frac{\partial^2 u}{\partial \zeta^2}$$

$$\frac{u}{h_1}\frac{\partial v}{\partial \xi} + \frac{v}{h_2}\frac{\partial v}{\partial \eta} + w\frac{\partial v}{\partial \zeta} - \frac{u^2}{h_1 h_2}\frac{\partial h_1}{\partial \eta} + \frac{uv}{h_1 h_2}\frac{\partial h_2}{\partial \xi} = \frac{-1}{h_2}\frac{\partial p}{\partial \eta} + \frac{\partial^2 v}{\partial \zeta^2}$$

$$\frac{1}{h_1 h_2}\left(\frac{\partial}{\partial \xi}(h_2 u) + \frac{\partial}{\partial \eta}(h_1 v)\right) + \frac{\partial w}{\partial \zeta} = 0$$

where u, v, w are the velocity components in the ξ, η, ζ directions and

$$h_1 = \left|\frac{\partial \mathbf{r}}{\partial \xi}\right|_{\zeta=0}, \quad h_2 = \left|\frac{\partial \mathbf{r}}{\partial \eta}\right|_{\zeta=0}.$$

Show that we can retrieve the three dimensional version of Prandtl's equations whenever the surface is developable (i.e. $h_1 = h_2 = 1$) and deduce that equations (2.23) are valid in the boundary layer on any two dimensional flow past a cylinder. If the streamlines for inviscid flow over a flat plate are **curved**, show that there is a secondary flow (i.e. a nonzero component of flow in a direction parallel to the plate but perpendicular to the free stream) in the boundary layer (Rosenhead p. 456).

3

Slow viscous flow

3.1 Flow at low Reynolds number

In this chapter we consider flows for which the Reynolds number is small. This is true in many practical situations (exercise 1) and we therefore consider (1.26) when $Re \ll 1$. We have already noted that the scaling used for the pressure in (1.26), which was chosen to make the pressure and inertia terms balance, is not now appropriate and that we need to rescale the dimensionless pressure with $\frac{1}{Re}$ so that it balances the dominant viscous terms. Replacing p by $\frac{1}{Re}p$ the equations become

$$\left.\begin{aligned} Re\left(\frac{\partial \mathbf{u}}{\partial t} + (\mathbf{u}.\nabla)\mathbf{u}\right) &= -\nabla p + \nabla^2 \mathbf{u} \\ \nabla.\mathbf{u} &= 0 \end{aligned}\right\}. \tag{3.1}$$

Again our strategy is to expand in powers of the small parameter Re. Usually we will be content to take only the first term in the expansion and we therefore write down the *slow flow* equations (also known as the equations of *creeping flow*) as

$$\nabla^2 \mathbf{u} = \nabla p \tag{3.2a}$$

and

$$\nabla.\mathbf{u} = 0. \tag{3.2b}$$

Since we are *not* neglecting the highest order derivatives, the solution to these equations represents the first term in a *regular* asymptotic expansion[1] in Re and we may expect it to be a reliable approximation as long as $Re \ll 1$. However there may be difficulties associated with such an expansion and we illustrate this by means of a simple ordinary differential equation example.

[1] See Appendix A.

3.2 A simple model equation

The equation we consider is

$$\frac{d^2u}{dx^2} + \varepsilon\frac{du}{dx} = 0$$

with $u(0) = 0$, $u(\infty) = 1$ and $0 < \varepsilon \ll 1$. We look for a solution as an asymptotic expansion in ε by writing $u \simeq u_0(x) + \varepsilon u_1(x) + \varepsilon^2 u_2(x) + \ldots$. Then, equating powers of ε yields a series of problems for $u_0, u_1, \ldots$. The first-order problem is

$$\frac{d^2u_0}{dx^2} = 0 \quad \text{with} \quad u_0(0) = 0 \quad \text{and} \quad u_0(\infty) = 1.$$

There is no solution to this problem and the best we can do is to satisfy $u_0(0) = 0$ by writing

$$u_0 = Cx \tag{3.3}$$

where C is some constant. In this simple case we can see what is really happening by solving the equation exactly to get

$$u = 1 - e^{-\varepsilon x}. \tag{3.4}$$

Thus the correct asymptotic expansion when $x = O(1)$ is

$$u = 0 + \varepsilon x - \frac{\varepsilon^2 x^2}{2} + \ldots.$$

This is only an asymptotic expansion as long as $\lim_{\substack{\varepsilon \to 0 \\ x \text{ fixed}}} \left|\frac{\varepsilon^2 u_2}{\varepsilon u_1}\right| = 0$. So the expansion 'breaks down' when $x = O(\varepsilon^{-1})$ and there is a 'boundary layer at infinity'. To find the solution in this 'outer region' we rescale by putting $x = \frac{X}{\varepsilon}$ so that the equation becomes $\frac{d^2u}{dX^2} + \frac{du}{dX} = 0$ with the condition $u(\infty) = 1$. The solution of this problem is $u = 1 - ke^{-X}$ where k is a constant and the only way we can 'match' this smoothly with the 'inner solution' (3.3) is to take $C = k - 1 = 0$. This clearly corresponds to the first-order approximation to (3.4) in each region. The need for great care to be taken with such problems is shown if we replace the condition $u(\infty) = 1$ by $u(L) = 1$ [Richardson]. Then we *can* obtain a solution that satisfies both boundary conditions, namely

$$u_0 = \frac{x}{L}, \quad u_1 = \frac{x}{2L}(L - x) \quad \text{etc.}$$

Now suppose we sought the value of $u'(0)$ when L was large. It would be tempting to say that, in this limit, u_0 is negligible and $u_1 \simeq \frac{x}{2}$, so $u'(0) = \frac{\varepsilon}{2}$. But, in fact, $u = \frac{1-e^{-\varepsilon x}}{1-e^{-\varepsilon L}}$ and $u'(0) = \varepsilon$ as $L \to \infty$!

This simple example serves to illustrate a new sort of difficulty that can occur in the asymptotic solution of a differential equation when the small parameter multiplies a low order derivative. This difficulty is caused by the boundary conditions being imposed at either end of a large range of the independent variable; no problem would have occurred if the condition $u = 1$ had been imposed at $x = L \ll O(\frac{1}{\varepsilon})$.

In the following two sections we consider uniform flow past a sphere and a circular cylinder and find that a solution to the slow flow equations (3.2) which satisfies the conditions both at the body and at infinity seems to be possible in the case of a sphere but not in the case of a cylinder.

3.3 Slow flow past a sphere

For a sphere in a uniform stream the Reynolds number is defined as $\frac{Ua}{\nu}$ where U is the velocity of the stream and a the radius of the sphere. When $Re \ll 1$, we need to consider the solution of (3.2) subject to the boundary conditions

$$\mathbf{u} = \mathbf{0} \quad \text{on} \quad r = 1 \quad \text{and} \quad \mathbf{u} \to \mathbf{i},\ p \to p_0 \quad \text{as} \quad r \to \infty. \tag{3.5}$$

Because the flow is axisymmetric we can work in terms of spherical polar coordinates (r, θ, ϕ) and then the continuity equation is

$$\frac{1}{r^2}\frac{\partial}{\partial r}(r^2 u_r) + \frac{1}{r\sin\theta}\frac{\partial}{\partial\theta}(\sin\theta u_\theta) = 0.$$

This allows us to introduce the *Stokes stream function* ψ which is defined by

$$u_r = \frac{1}{r^2\sin\theta}\frac{\partial\psi}{\partial\theta}, \quad u_\theta = -\frac{1}{r\sin\theta}\frac{\partial\psi}{\partial r}$$

or

$$\mathbf{u} = \mathrm{curl}(0, 0, \frac{\psi}{r\sin\theta}). \tag{3.6}$$

Since $\nabla^2\mathbf{u} = \text{grad div } \mathbf{u}$ - curl curl $\mathbf{u}$, equation (3.2*a*) can be rewritten as

$$\mathrm{curl}\ \mathrm{curl}\,\mathbf{u} = -\nabla p,$$

or, eliminating p by taking the curl of this equation,

$$\mathrm{curl}^3\mathbf{u} = \mathbf{0}. \tag{3.7}$$

Using (3.6) the equation for ψ is therefore

$$\text{curl}^4(0, 0, \frac{\psi}{r\sin\theta}) = \mathbf{0}. \tag{3.8}$$

Now

$$\text{curl}^2(0, 0, \frac{\psi}{r\sin\theta}) = (0, 0, \frac{-D^2\psi}{r\sin\theta})$$

where

$$D^2 = \frac{\partial^2}{\partial r^2} + \frac{1}{r^2}\frac{\partial^2}{\partial \theta^2} - \frac{\cot\theta}{r^2}\frac{\partial}{\partial \theta} \quad (\text{note } D^2 \neq \nabla^2!) \tag{3.9}$$

and so (3.8) reduces to

$$D^4\psi = 0. \tag{3.10}$$

The boundary conditions (3.5) can be written in terms of ψ as

$$\psi = \frac{\partial \psi}{\partial r} = 0 \quad \text{on} \quad r = 1$$

and (3.11)

$$\psi \sim \frac{1}{2}r^2 \sin^2\theta \quad \text{as} \quad r \to \infty .$$

By separating the variables we find that there is a suitable solution of (3.10) of the form $\psi = f(r)\sin^2\theta$. It can be shown easily that

$$D^2\psi = \left(\frac{d^2}{dr^2} - \frac{2}{r^2}\right) f \sin^2\theta$$

and hence

$$\left(\frac{d^2}{dr^2} - \frac{2}{r^2}\right)^2 f = 0.$$

The general solution of this "homogeneous" ordinary differential equation for $f(r)$ is $f = Ar^4 + Br^2 + Cr + Dr^{-1}$ and the boundary condition at infinity shows that $A = 0$ and $B = \frac{1}{2}$ while the condition on the sphere implies that $C = -\frac{3}{4}$ and $D = \frac{1}{4}$. This is the *Stokes solution* for flow past a sphere

$$\psi = \left(\frac{1}{2}r^2 - \frac{3}{4}r + \frac{1}{4}r^{-1}\right)\sin^2\theta, \tag{3.12}$$

which can be shown to be the unique solution to the problem posed in (3.10) and (3.11), assuming ψ is sufficiently smooth. The velocity components are

$$u_r = \left(1 - \frac{3}{2r} + \frac{1}{2r^3}\right)\cos\theta, \quad u_\theta = \left(-1 + \frac{3}{4r} + \frac{1}{4r^3}\right)\sin\theta \tag{3.13}$$

and the pressure, which can be calculated from equation (3.2), is

$$p = p_0 - \frac{3\cos\theta}{2r^2}.$$

This solution, although in an infinite domain, satisfies the boundary conditions both at infinity and on the body and we may suppose that it is a uniformly valid approximation. In fact it can be checked that the neglected term, $Re(\mathbf{u}.\nabla)\mathbf{u}$, is everywhere small compared to the other terms and so this supposition is correct.

We now compute the drag on the sphere; this is a quantity of great practical interest.[2] It results from both the normal and tangential stresses on the sphere and is given nondimensionally by

$$\int\int_{|r|=1} (\sigma_{rr}\cos\theta - \sigma_{r\theta}\sin\theta)_{r=1} dS. \qquad (3.14)$$

Now, in spherical polar coordinates

$$\sigma_{rr} = -p + 2\frac{\partial u_r}{\partial r}$$

and

$$\sigma_{r\theta} = r\frac{\partial}{\partial r}\left(\frac{u_\theta}{r}\right) + \frac{1}{r}\frac{\partial u_r}{\partial \theta},$$

so, using (3.13) and (3.14) to evaluate σ_{rr} and $\sigma_{r\theta}$ on the sphere, we get

$$\sigma_{rr}\,|_{r=1} = -p_0 + \frac{3\cos\theta}{2} \quad \text{and} \quad \sigma_{r\theta}\,|_{r=1} = -\frac{3}{2}\sin\theta.$$

The contribution to the drag from the normal stresses is

$$\int\int \sigma_{rr}\cos\theta\, dS = 2\pi\int_0^\pi \left(-p_0 + \frac{3\cos\theta}{2}\right)\cos\theta\sin\theta d\theta = 2\pi$$

and the contribution from the shear stresses is

$$\int\int -\sigma_{r\theta}\sin\theta dS = 2\pi\int_0^\pi \frac{3}{2}\sin^2\theta\sin\theta d\theta = 4\pi$$

Hence the contribution from the shear stresses is twice that due to normal stresses and the total drag in dimensional terms is $6\pi\mu Ua$. This famous *Stokes formula* is typical of forces exerted on bodies in slow viscous flow in that it is linear in the dynamic viscosity μ, the free stream velocity U, and a typical body dimension a.

[2] For example, one application of this result is used in Millikan's oil drop experiment to determine the charge on an electron.

3.4 Slow flow past a circular cylinder

In two dimensions we can introduce a stream function by writing

$$\mathbf{u} = \left(\frac{\partial \psi}{\partial y}, -\frac{\partial \psi}{\partial x}, 0\right) = \text{curl}(0, 0, \psi)$$

where we are now working in Cartesian coordinates. From (3.7), we see that in slow flow the stream function ψ satisfies

$$\text{curl}^4(0, 0, \psi) = \mathbf{0}$$

or, since $\text{curl}^2(\psi\mathbf{k}) = -\nabla^2\psi\mathbf{k}$,

$$\nabla^4\psi = 0 \tag{3.15}$$

which is called the *biharmonic equation.* Not surprisingly, since it is the "square" of Laplace's equation,[3] this equation can be solved by complex variable methods (see exercise 5) but we do not need such elaborate machinery when we concentrate on the specific example of flow past a circular cylinder in a uniform stream. The boundary conditions for this example are, in plane polar coordinates

$$\psi = \frac{\partial \psi}{\partial r} = 0 \quad \text{on} \quad r = 1 \tag{3.16}$$

and

$$\psi \sim y = r\sin\theta \quad \text{as} \quad r \to \infty \tag{3.17}$$

where, as before, we have taken the free stream velocity and the radius of the cylinder as our reference quantities. Separating the variables, we find that $\psi = f(r)\sin\theta$ is a suitable solution if

$$\left(\frac{d^2}{dr^2} + \frac{1}{r}\frac{d}{dr} - \frac{1}{r^2}\right)^2 f = 0 \tag{3.18}$$

with $f(1) = f'(1) = 0$ and $f \to r$ as $r \to \infty$. Now $f = Ar^n$ is a solution of (3.18) if $n = 3, 1, 1$ or -1 and so the general solution is

$$f = Ar^3 + Br\log r + Cr + Dr^{-1}. \tag{3.19}$$

In order to satisfy the condition at infinity, both A and B must vanish and $C = 1$ but then we cannot satisfy both boundary conditions at $r = 1$. This is an example of the Whitehead paradox; it can be proved that there is no solution to the problem posed in (3.15)-(3.17) and our failure to find

[3] This equation also arises in many "two-dimensional" problems in elasticity, such as the bending of plates. This enables numerous analogies to be drawn between elastic displacements and slow fluid velocities.

a solution is not a result of assuming a particular form for ψ. Further, it can be shown that there is no solution to the problem of uniform slow flow past a cylinder of arbitrary cross-section. This result led to speculation, for over 70 years, that steady flow at low Reynolds number past a cylinder was unattainable. In around 1930, Oseen suggested that the creeping flow model (3.2) should be replaced by the ad hoc *Oseen* model

$$Re\frac{\partial \mathbf{u}}{\partial x} = -\nabla p + \nabla^2\mathbf{u} \qquad (3.20a)$$

$$\nabla.\mathbf{u} = 0 \qquad (3.20b)$$

in which the awkward nonlinear term in the Navier-Stokes equations is replaced by a linear term which is a good approximation at large distances from the cylinder. Indeed, Oseen was able to solve (3.20) with $\mathbf{u} \to \mathbf{i}$ at ∞ but he was unable to justify the retention of the apparently small left-hand side of (3.20a). It was not until 1957 that it was understood that the regular asymptotic expansion in powers of Re of the solution to (3.1) is only valid when $r \ll O(\frac{1}{Re})$ [Proudman and Pearson]. When $r = O(\frac{1}{Re})$ it is necessary to rescale $\mathbf{x} = \frac{1}{Re}\mathbf{x}'$ in (3.1). This would lead to the full Navier-Stokes equations if it were not for the fact that far from the cylinder the flow is close to the free stream velocity so it is possible to write $\mathbf{u} \simeq \mathbf{i} + \varepsilon\mathbf{u}' + \ldots$ where ε is a small parameter that depends on Re and is determined in exercise 8. Then (3.1) becomes, to the first order in ε and Re,

$$\frac{\partial \mathbf{u}'}{\partial x'} = -\nabla' p' + \nabla'^2\mathbf{u}' \qquad (3.21)$$

where we have scaled $p = \varepsilon Re p'$ in order to balance the three terms. This is equivalent to Oseen's equation (3.20) and we can now see how the solutions fit together because the situation is analogous to the situation described in section 3.2. The slow flow model (3.2) satisfying conditions (3.16) on the cylinder is an 'inner' solution valid for $r \ll O(\frac{1}{Re})$ but for $r \geq O(\frac{1}{Re})$, Oseen's equation (3.21) gives the correct first approximation to the 'outer' solution. The boundary condition (3.17) at infinity is applied to the Oseen solution and then these inner and outer solutions need to be 'matched' together to determine the full solution for all r. The reason that this took so long to understand, and that the full solution and matching were only completed in 1957, was because of complications caused by the log term in (3.19); in fact the inner solution has to be constructed in the form $\psi \simeq \dfrac{\psi_0}{\log 1/Re} + \psi_1 + \ldots$. These complications are the reason that we relegate the details of this solution to exercise 8.

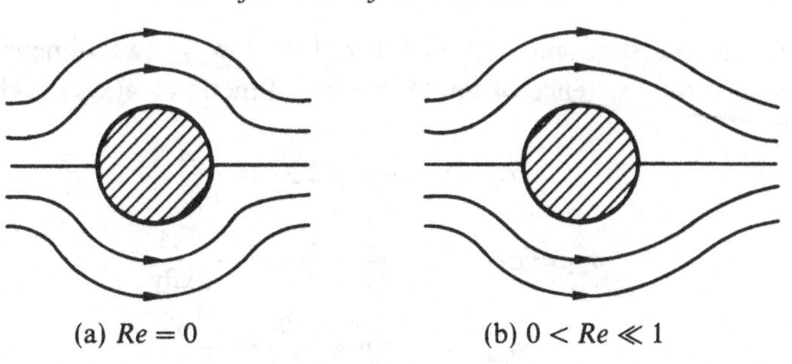

Fig. 3.1. Slow flow past a sphere

It is worth noting that although this difficulty did not occur *to first order* for the sphere problem considered in section 3.3, if we had continued with the asymptotic expansion in powers of *Re*, it would have broken down at the *second* term. The drag on the sphere is in fact

$$6\pi\mu Ua(1+\frac{3}{8}Re+\frac{9}{40}Re^2\log Re+\frac{1}{40}(9\gamma+15\log 2-\frac{323}{40})Re^2+\ldots) \quad (3.22)$$

where γ is Euler's constant and the third and fourth terms can only be evaluated by matching an inner and an outer expansion.

We also remark that it is only when we solve Oseen's equation (3.20) that we detect any fore-and-aft asymmetry in the streamline pattern for flow past a sphere or cylinder (see exercise 10 and figure 3.1b). Indeed, the streamlines for the zero Reynolds number flow corresponding to (3.13) are symmetric (figure 3.1a) (as in irrotational flow past a sphere), but *steady* asymmetric flows *can* be observed experimentally unlike the wake flows described in Chapter 2. However even in relatively slow flow, oscillatory instabilities develop at Reynolds numbers greater than a certain critical value.

3.5 Slow flow with free boundaries

As with surface gravity waves, where the field equation for the potential is just Laplace's equation, the presence of a free boundary in any fluid mechanics problem causes almost insuperable nonlinearity. This is especially true in steady slow viscous flow; in two dimensions, it is easy enough to impose the boundary condition $\psi = 0$ on the biharmonic equation but the stress free conditions $\boldsymbol{\sigma}.\mathbf{n} = \mathbf{0}$ are very difficult to manage. However the following ideas can be of some help.

We can note that since $\frac{\partial \sigma_{ij}}{\partial x_j} = 0$ for slow flow, in two dimensions we can deduce the existence of an "Airy stress function" $A(x, y)$ such that

$$\begin{aligned} \sigma_{11} &= -p + 2\frac{\partial u}{\partial x} = \frac{\partial^2 A}{\partial y^2} \\ \sigma_{21} = \sigma_{12} &= \frac{\partial u}{\partial y} + \frac{\partial v}{\partial x} = -\frac{\partial^2 A}{\partial x \partial y} \\ \sigma_{22} &= -p + 2\frac{\partial v}{\partial y} = \frac{\partial^2 A}{\partial x^2}. \end{aligned}$$

Like the stream function ψ, which is only determined to within a constant, A is only determined to within an arbitrary linear function of x, y (in physics jargon,ψ and A are called "gauges"). We also note that $p = -\frac{1}{2}\nabla^2 A$ and, since $\nabla^2 p = 0$ in any slow flow, A is a biharmonic function. However, the real advantage of using A is apparent when we write the free boundary as $x = x(s)$, $y = y(s)$ and see that $\boldsymbol{\sigma}.\mathbf{n} = \frac{\partial}{\partial s}\left(-\frac{\partial A}{\partial y}, \frac{\partial A}{\partial x}\right)$ so that ∇A is constant along the free boundary. Given our flexibility in the choice of A, we can hence take $A = \frac{\partial A}{\partial n} = 0$ on the boundary. But the price we have to pay for working with A is that the kinematic condition $\psi = 0$ is now unmanageable!

The intricate relationships between A, ψ and p can best be summarised in terms of the complex variable formulation. The solution of $\nabla^4 \psi = 0$ is (exercise 5)

$$\psi = Rl(\bar{z}\phi(z) + \chi(z))$$

from which it follows that

$$A = Rl(-2i\bar{z}\phi(z) - 2i\chi(z))$$

and hence on a free boundary $z = z(s)$ where $A = 0$ and $\nabla A = \mathbf{0}$,

$$\bar{z}\phi(z) + \chi(z) = 0$$

and

$$\bar{z}\phi'(z) + \chi'(z) - \overline{\phi(z)} = 0$$

which leads finally to the boundary conditions

$$\bar{z}\phi(z) + \chi(z) = 0 \quad \text{and} \quad Rl\left(\phi(z)\frac{d\bar{z}}{ds}\right) = 0. \tag{3.23}$$

It is even possible to incorporate surface tension into this formulation without too much difficulty.

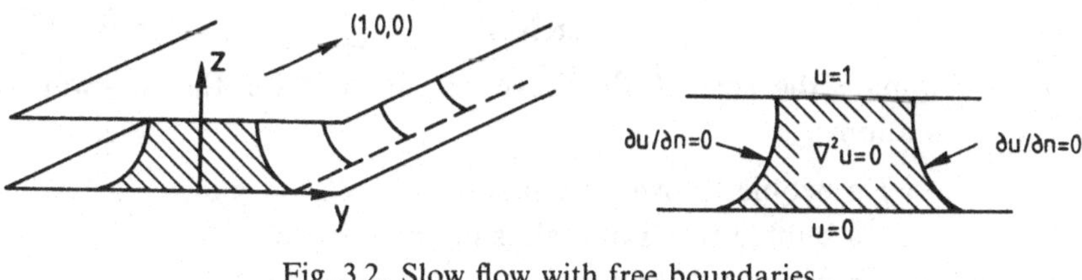

Fig. 3.2. Slow flow with free boundaries

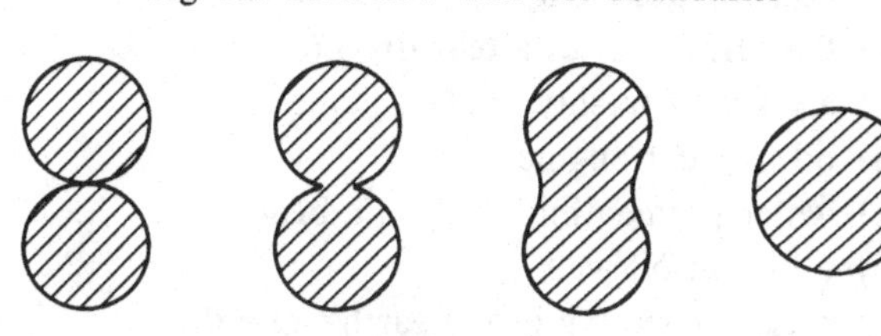

Fig. 3.3. Sintering of two fibres

We conclude by mentioning that, because we have neglected inertia, the *evolution* of all slow flows is indeterminate unless there is a time derivative, say, in a kinematic free boundary condition. Indeed we can always add any arbitrary time-dependent rigid body motion that is compatible with the boundary conditions. The possibilities are increased when there are free boundaries involved, as for example, when a fluid is undergoing Couette flow but is unconstrained laterally as in figure 3.2. In this geometry, we see that the slow flow equation, the kinematic condition, and the mechanical boundary conditions are all satisfied if $\mathbf{u} = (u(y, z), 0, 0)$ where $\nabla^2 u = 0$ and $u = 1$ on $z = 1$, $u = 0$ on $z = 0$, $\frac{\partial u}{\partial n} = 0$ on the free boundary. Even when the total mass flow is prescribed, this clearly has a solution whatever the shape of the free boundary. Thus we may expect that other agencies such as gravity, surface tension, or inertia may exert a decisive influence on such problems, even though the associated dimensionless parameter is very small. A typical example is the sintering of two glass fibres which, when they are placed in contact along a generator, evolves as shown in figure 3.3 as long as surface tension (however small) is included; without surface tension, there is no evolution and any of the situations in figure 3.3 (or those in between) is a possible equilibrium state.

Exercises

1 Estimate the size of the Reynolds number for the following situations:

(i) a pebble thrown in a pond
(ii) a bubble rising in a glass of champagne
(iii) pouring golden syrup
(iv) the formation of a tear drop (crying)
(v) a layer of freshly applied paint
(vi) oil in a car engine
(vii) water permeating sand on a beach
(viii) oil in an oil well
(ix) rock convecting in the earth's mantle.

The kinematic viscosity of the following fluids at normal temperatures is, to within a factor of 10,

water, tear fluid	$10^{-6}\text{m}^2/\text{s}$
air	$10^{-5}\text{m}^2/\text{s}$
engine oil	$10^{-4}\ \text{m}^2/\text{s}$
paint	$10^{-3}\ \text{m}^2/\text{s}$
crude oil	$10^{-5}\ \text{m}^2/\text{s}$
syrup	$10^{-1}\ \text{m}^2/\text{s}$
mantle rock	$10^{18}\ \text{m}^2/\text{s}$.

2 When Re is a *small* positive parameter, show how the solution of the ordinary differential equation

$$x\frac{dy}{dx} + y = Re\, y^{\frac{1}{2}}, \quad x \geq 1$$

with $y(1) = 1$ can be found as an expression in powers of Re. Evaluate the first two terms in this expression and explain why it is not valid as $x \to \infty$.

3 A sphere of radius a is placed in a uniform stream of velocity U and viscosity ν where $\frac{Ua}{\nu} \ll 1$. Starting from (3.12), show that on $\theta = 0$, the dimensional velocity is approximately $U\left(1 - \frac{3a}{2r}\right)$ when $\frac{a}{r} \ll 1$. Given that the dimensional drag on the sphere is $6\pi\mu Ua$, deduce that, when Re is negligible, the drag on a second sphere of radius b inserted in the same flow with its centre at $\theta = 0$, $r = l$ will be approximately $6\pi\mu Ub\left(1 - \frac{3a}{2l}\right)$ when $l \gg a$ and b. Explain the significance of the second of these two assumptions.

4 (i) Surface tension forces are strong enough that a bubble rising in a viscous fluid is spherical and of radius a. Assume the surface tension is constant and there is negligible tangential stress on the fluid at the bubble surface. Deduce that it will rise with a velocity U given approximately by $\frac{1}{3}\frac{ga^2}{\nu}$ provided $\frac{aU}{\nu} \ll 1$.
(ii) Show that for a general shape of convex bubble the boundary conditions at the surface are

$$\sigma_{nn} = T\kappa, \; \sigma_{ns} = \frac{\partial T}{\partial s}$$

where κ is the **mean** curvature and T is the surace tension.

5 (i) Verify that in two-dimensional slow flow the solution of the biharmonic equation

$$\nabla^4 \psi = 0$$

may be written in the form

$$\psi = \text{Rl}(\bar{z}\phi(z) + \chi(z))$$

when ϕ, χ are holomorphic functions of $z = x + iy$. Deduce that the velocity components u, v are given by

$$v + iu = -(\overline{\phi(z)} + \bar{z}\phi'(z) + \chi'(z)).$$

Fluid is contained in the positive quadrant of the x, y plane between rigid walls at $x = 0$ and $y = 0$. There is a sink of magnitude Q at the origin. By taking $\phi = \frac{c}{z}$ and $\chi = 0$, find a suitable solution for the stream function ψ. Is this solution unique?
(ii) Show that, if surface tension effects are taken into account, the boundary conditions (3.23) on the free surface become

$$\bar{z}\phi(z) + \chi(z) = 0 \quad \text{and} \quad \frac{T}{\mu}\left|\frac{dz}{ds}\right|^2 = \pm 4\text{Rl}\left(\phi(z)\frac{d\bar{z}}{ds}\right)$$

where the sign taken depends on whether the boundary is concave or convex with respect to the fluid.

*6 By considering $\phi(z) = z^\lambda$ and $\chi(z) = z^{\lambda+1}$ in the formula for ψ in exercise 5, show that ψ can represent symmetric flow in a wedge $-\alpha < \theta < \alpha$ provided that $(\lambda - 1)\tan(\lambda - 1)\alpha = (\lambda + 1)\tan(\lambda + 1)\alpha$. Given that this equation has complex roots for λ, deduce that the wall shear stress vanishes at an infinite number of points in any neighbourhood of $r = 0$. Sketch a streamline pattern consistent with this result.

7 Two-dimensional steady flow is generated by a source or sink of strength M at the vertex of wedge of angle 2α. If $\frac{M}{\alpha\nu} \ll 1$, show that the similarity solution $\psi = f(\theta)$ approximately satisfies

$$f'''' + 4f'' = 0 \quad \text{with} \quad f'(\pm\alpha) = 0$$

(see exercise 8, Chapter 2). Determine f and sketch the flow field for several values of α. Show that there is a singularity where $\tan 2\alpha = 2\alpha$. This breakdown of a similarity solution at a critical value of a parameter is not uncommon and its resolution has only recently been understood. It is associated with the emergence of eigensolutions at these critical parameter values.

*8 Construct a uniformly valid solution for steady flow past a circular cylinder with $Re = \frac{Ua}{\nu} \ll 1$ by the following steps.
(i) Show that there is an 'inner' solution, which satisfies the slow flow equations (3.2) and the boundary conditions on the cylinder, of the form

$$\psi_i = B(r \log r - \frac{1}{2}r + \frac{1}{2r}) \sin\theta,$$

where B may depend on Re.
(ii) Show, by writing $r = \frac{1}{Re}\hat{r}$, that

$$\psi_i \sim \frac{B}{Re} \log\left(\frac{1}{Re}\right) \hat{r} \sin\theta$$

when $r = O\left(\frac{1}{Re}\right)$.
(iii) Show by writing $\psi = \frac{1}{Re}\hat{\psi}$ that the 'outer' problem is

$$\frac{\partial(\hat{\psi}, \hat{\nabla}^2\hat{\psi})}{\partial(\hat{y}, \hat{x})} = \hat{\nabla}^4\hat{\psi}$$

with $\hat{\psi} \sim B \log\left(\frac{1}{Re}\right)\hat{y}$ as $\hat{r} \to 0$ and $\hat{\psi} \to \hat{y}$ as $\hat{r} \to \infty$. Deduce that $B = \dfrac{1}{\log\left(\frac{1}{Re}\right)}$.
(iv) By expanding $\hat{\psi} = \hat{y} + \frac{1}{\log\frac{1}{Re}}\hat{\psi}_1 + \ldots$ derive the equation

$$\frac{\partial}{\partial \hat{x}}\left(\hat{\nabla}^2\hat{\psi}_1\right) = \hat{\nabla}^4\hat{\psi}_1$$

and show that it is equivalent to Oseen's equation (3.20) in the outer region. What are the appropriate boundary conditions on $\hat{\psi}_1$?

*9 Use the solution constructed in exercise 8 to show that the drag per unit length on a circular cylinder in slow flow is approximately $\dfrac{4\pi\mu U}{\log\left(\frac{\nu}{Ua}\right)}$.

[In dimensionless terms the stress can be expressed in plane polar coordinates as $\sigma_{rr} = -p + 2\frac{\partial u_r}{\partial r}$ and $\sigma_{r\theta} = r\frac{\partial}{\partial r}\left(\frac{u_\theta}{r}\right) + \frac{1}{r}\frac{\partial u_r}{\partial \theta}$.]

*10 Show that Oseen's equation (3.20) can be written as

$$\left(\frac{1}{Re}D^2 - \cos\theta\frac{\partial}{\partial r} + \frac{\sin\theta}{r}\frac{\partial}{\partial \theta}\right)D^2\psi = 0 \qquad (*)$$

in spherical polar coordinates, where D^2 is defined in (3.9).
(i) By writing $\cos\theta = c$, show that (*) is

$$\left(\frac{1}{Re}D^2 - c\frac{\partial}{\partial r} - \frac{1-c^2}{r}\frac{\partial}{\partial c}\right)D^2\psi = 0 \qquad (**)$$

(ii) Verify that $\psi = (1+c)\left[1 - e^{-\frac{1}{2}Re\,r(1-c)}\right]$ satisfies (**).
(iii) Verify that

$$\psi = \left(\frac{r^2}{2} + \frac{1}{4r}\right)\sin^2\theta - \frac{3}{2Re}(1+c)\left[1 - e^{-\frac{1}{2}Re\,r(1-c)}\right]$$

also satisfies (**) and the conditions (3.11) when terms of $O(Re^2)$ are neglected.
(iv) Deduce that the streamlines corresponding to this stream function only give an appreciable correction to the streamlines of (3.12) in a wake where $r\theta^2 = O(1/Re)$.

4

Thin films

In this chapter we consider relatively low Reynolds number flow of a thin film. Such a film may exist between two rigid walls, as in a bearing, or in a droplet, e.g. paint, spreading under gravity on a rigid surface. In either case the geometry of the problem allows us to simplify equation (3.2) in a way that is similar to the technique used to derive boundary layer theory in Chapter 2. The differences are that the order of magnitude of the width of the thin layer is dictated by the data of the problem and, since the layer is confined geometrically, there is no need to match with an outer flow.

4.1 Lubrication theory for slider bearings

The simple observation that a sheet of paper can slide across a smooth floor shows that a thin layer of fluid can support a relatively large normal load while offering very little resistance to tangential motion. More important mechanical examples occur in the lubrication of machinery and this motivates the study of *slider bearings.* A slider bearing consists of a thin layer of viscous fluid confined between nearly parallel walls that are in relative tangential motion.

A two-dimensional bearing is shown in figure 4.1 in which the plane $y = 0$ moves with constant velocity U in the x-direction and the top of the bearing (the slider) is fixed. The variables are nondimensionalised with respect to U and the length L of the bearing so that the position of the slider is given in the dimensionless variables used in (3.1) by $y = \delta H(x)$ where δL is the typical gap-width of the bearing. The basic assumption of lubrication theory is that $\delta \ll 1$ so that we can use the ideas of boundary layer theory to simplify the Navier-Stokes equations. Starting from the steady form of (3.1), we rescale y, v by writing $y = \delta y'$,

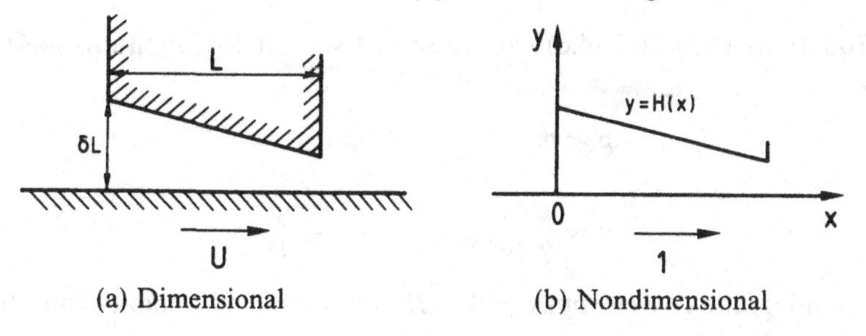

(a) Dimensional (b) Nondimensional

Fig. 4.1. Slider bearing

$v = \delta v'$ to get

$$Re\left(u\frac{\partial u}{\partial x} + v'\frac{\partial u}{\partial y'}\right) = -\frac{\partial p}{\partial x} + \frac{\partial^2 u}{\partial x^2} + \frac{1}{\delta^2}\frac{\partial^2 u}{\partial y'^2} \tag{4.1}$$

$$\delta Re\left(u\frac{\partial v'}{\partial x} + v'\frac{\partial v'}{\partial y'}\right) = -\frac{1}{\delta}\frac{\partial p}{\partial y'} + \delta\frac{\partial^2 v'}{\partial x^2} + \frac{1}{\delta}\frac{\partial^2 v'}{\partial y'^2} \tag{4.2}$$

$$\frac{\partial u}{\partial x} + \frac{\partial v'}{\partial y'} = 0 \tag{4.3}$$

with boundary conditions

$$\left.\begin{array}{lll} u = 1,\ v' = 0 & \text{on} & y' = 0 \\ u = 0,\ v' = 0 & \text{on} & y' = H(x) \end{array}\right\}. \tag{4.4}$$

We also need boundary conditions at $x = 0, 1$ but we will return to them later. The only way that (4.1) will not reduce to a triviality as $\delta \to 0$ is if the pressure is rescaled with $\frac{1}{\delta^2}$. Thus we write $p = \frac{1}{\delta^2}p'$ and, to lowest order, the equations are (on dropping dashes)

$$0 = -\frac{\partial p}{\partial x} + \frac{\partial^2 u}{\partial y^2}, \tag{4.5}$$

$$0 = \frac{\partial p}{\partial y}, \tag{4.6}$$

and

$$\frac{\partial u}{\partial x} + \frac{\partial v}{\partial y} = 0. \tag{4.7}$$

These equations are the *lubrication model* and are based on the two assumptions that $\delta \ll 1$ *and* $Re\delta^2 \ll 1$. Notice that it is not necessary for the Reynolds number based on L to be small but only that the *reduced Reynolds number* $Re\delta^2 = \frac{UL\delta^2}{\nu}$ be small.

Equations (4.5) and (4.6) can be solved subject to conditions (4.4) to give

$$p = p(x)$$
$$u = \frac{1}{2}\frac{dp}{dx}y(y-H) + 1 - \frac{y}{H}. \tag{4.8}$$

Then, integrating (4.7) from $y = 0$ to H and using the boundary condition (4.4) on v leads to

$$\int_0^{H(x)} \left(\frac{\partial u}{\partial x}\right) dy = 0.$$

Thus

$$\frac{d}{dx}\int_0^{H(x)} u dy = \int_0^{H(x)} \frac{\partial u}{\partial x} dy + H'(x)u\Big|_{y=H} = 0$$

and so, substituting for u from (4.8),

$$\frac{1}{6}\frac{d}{dx}\left(H^3\frac{dp}{dx}\right) = \frac{dH}{dx}. \tag{4.9}$$

This is *Reynolds equation* and, given $H(x)$, it can be solved for p if the value of the pressure is known at the ends of the bearing. Before solving this equation we note that the *dimensional* stress tensor is

$$\boldsymbol{\sigma} = \frac{\mu U}{L}\begin{bmatrix} -\frac{1}{\delta^2}p + 2\frac{\partial u}{\partial x} & \frac{1}{\delta}\frac{\partial u}{\partial y} + \delta\frac{\partial v}{\partial x} \\ \frac{1}{\delta}\frac{\partial u}{\partial y} + \delta\frac{\partial v}{\partial x} & -\frac{1}{\delta^2}p + 2\frac{\partial v}{\partial y} \end{bmatrix}. \tag{4.10}$$

The stress T_i, exerted on the upper surface is $\sigma_{ij}n_j$ where the normal $\mathbf{n}$ is given by $(\delta H', -1)/(1+\delta^2 H'^2)^{1/2}$ and so, to lowest order in δ,

$$\mathbf{T} = \frac{\mu U}{L}\left[\frac{1}{\delta}\left(-H'p - \frac{\partial u}{\partial y}\right), \frac{1}{\delta^2}p\right]. \tag{4.11}$$

The normal stress exerted on the upper surface is therefore an order of magnitude greater than the tangential stress. It is interesting to note that it has recently been proposed to use air bearings in aeroplane engines in which a 20 μm lubricating air layer at the end of the shaft in the jet engine will support a load of 10^4kg. This is possible by taking δ to be very small (and using some very accurate engineering to prevent contact!).

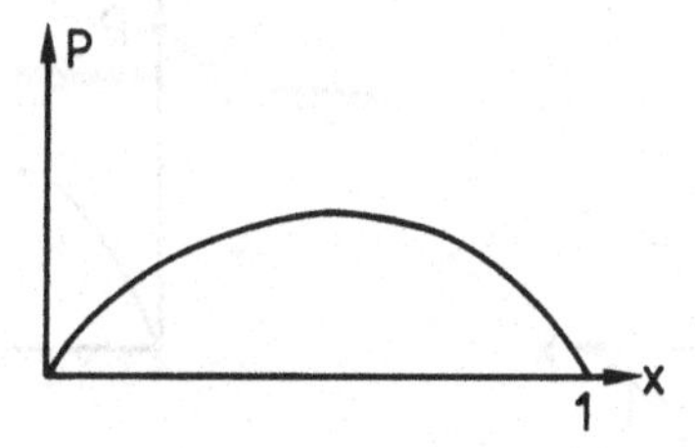

Fig. 4.2. Pressure in a converging bearing ($H = 1 - \frac{x}{4}$)

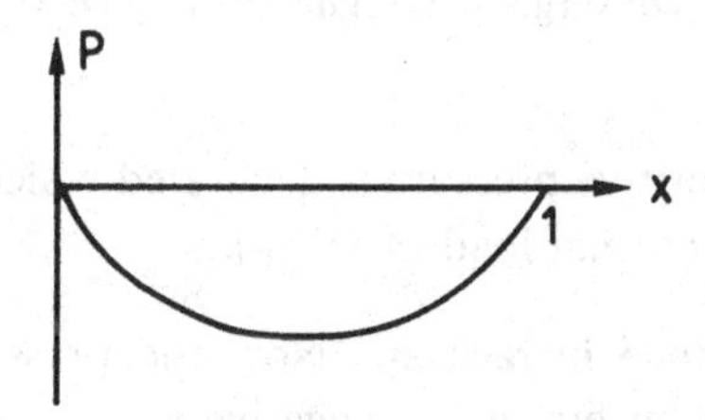

Fig. 4.3. Pressure in a diverging bearing ($H = 1 + \frac{x}{4}$)

We now complete the solution of Reynolds equation (4.9) by integrating it once to give

$$\frac{1}{6}H^3\frac{dp}{dx} = H - H_0, \quad H_0 = \text{const.}$$

If we assume that the ends of the bearing are at ambient pressure $p = 0$,[1] we get

$$p(x) = 6\int_0^x \frac{H(\bar{x}) - H_0}{H^3(\bar{x})}d\bar{x} \tag{4.12}$$

where the constant H_0 is determined by $\int_0^1 \frac{H(\bar{x})-H_0}{H^3(\bar{x})}d\bar{x} = 0$.

We now consider this solution in various different bearing geometries.

1. $H(x)$ **monotone decreasing.** For the linear bearing shown in figure 4.1, the pressure can be calculated from (4.12) (exercise 1). As shown in figure 4.2, the pressure is everywhere positive with a maximum value at x_m where $H(x_m) = H_0$. Thus when lubricant is forced into a *converging*

[1] The scaling that we have used for the pressure refers to the pressure *variations*. We may therefore write the dimensional pressure as $p_0 + \frac{\mu U}{L\delta^2}p$ where p_0 is the ambient pressure. Also in practice, there will be 'fully' two-dimensional flow near the ends of the bearing where x and y derivatives become comparable. However, as long as $Re \ll 1$, a conservation argument can be used on the slow flow equations which shows that pressure variations across these end regions are small and we can therefore apply the pressure boundary condition directly to solutions of the lubrication equations.

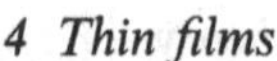

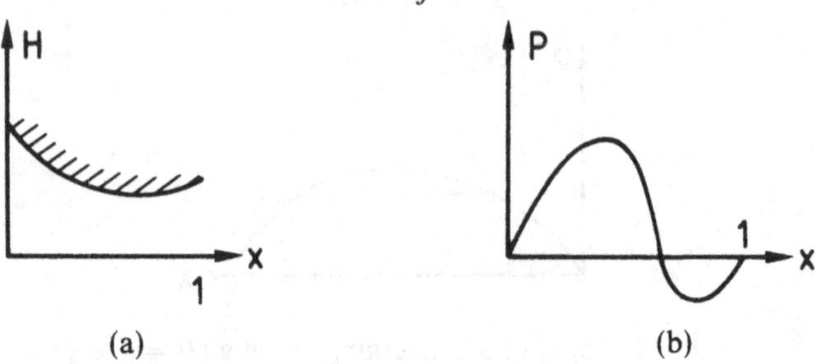

Fig. 4.4. A converging-diverging bearing ($H = 1 - \frac{3}{2}x + \frac{9}{8}x^2$)

bearing, a large positive pressure is generated which, from (4.11), can be seen to support a normal load of $O(\frac{\mu U}{\delta^2})$.

2. $H(x)$ **monotone increasing.** Now the pressure determined from (4.12) is, as shown in figure 4.3, negative everywhere. This suggests the possibility of *cavitation* if the dimensional pressure becomes sufficiently small. We will discuss this further at the end of this section.

3. $H(x)$ **nonmonotone.** If $H(x)$ decreases and then increases, as in figure 4.4, the pressure distribution in the bearing will be first positive and then negative. Such a bearing will be able to support a load if $\int_0^1 p dx > 0$ but there is again the possibility of cavitation. This configuration is relevant in the commonly occurring *journal bearing* (figure 4.5) where a rotating shaft is supported by a thin film of fluid within a surrounding journal. Reynolds' equation can be formulated in cylindrical polar coordinates (exercise 2) but as long as the gap is small compared with the radius of the shaft the equations will reduce to (4.5)-(4.7). The crucial region in a journal bearing is where the gap is narrowest (near A in figure 4.5) and here, as discussed above, the pressure will be positive to the left of A ($\theta < 0$) and will become negative to the right of A where $\theta > 0$. Thus the dominant force on the rotating shaft is *perpendicular* to the line OA. This can cause the (undesirable) phenomenon of 'whirl' instability which causes the shaft to orbit around the centre of the journal in a possibly chaotic vibration.

There is also the possibility of cavitation in $\theta > 0$ which may mean that *contact* and *wear* can occur in the bearing. Indeed cavitation is a primary source of wear in all mechanical bearings and its modelling is an important but difficult problem. The simplest model is to 'truncate' p at the cavitation pressure p_c so that the pressure is as shown in figure 4.6. But it is more realistic to solve the Reynolds equation with a 'free

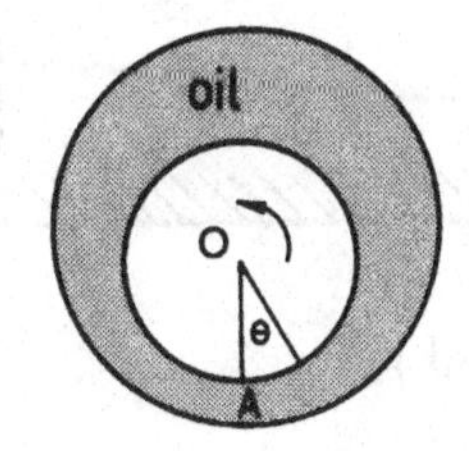

Fig. 4.5. A journal bearing

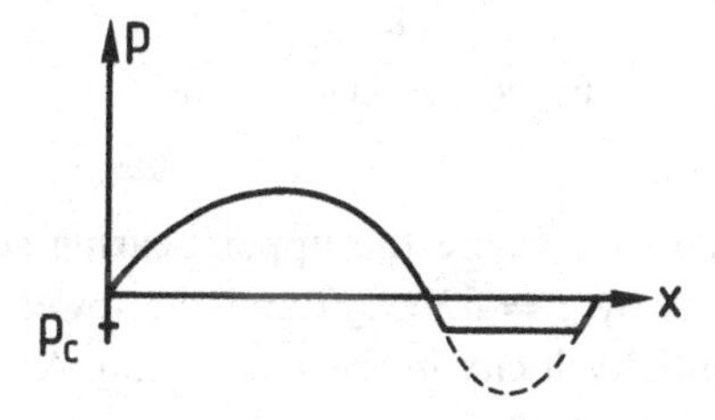

Fig. 4.6. Simple cavitation model

boundary' (as shown in figure 4.7) on which $p = p_c$. The mass flow or 'leakage' around the cavity is proportional to $\frac{\partial p}{\partial \theta}$ and this must be postulated at the ends of the cavity. The position of the ends of the cavity (θ_1 and θ_2) has to be determined as part of the problem. The easiest case is to assume *no* leakage and then a variational formulation of the problem is possible (exercise 11).

The derivation of Reynolds equation (4.9) can easily be extended to

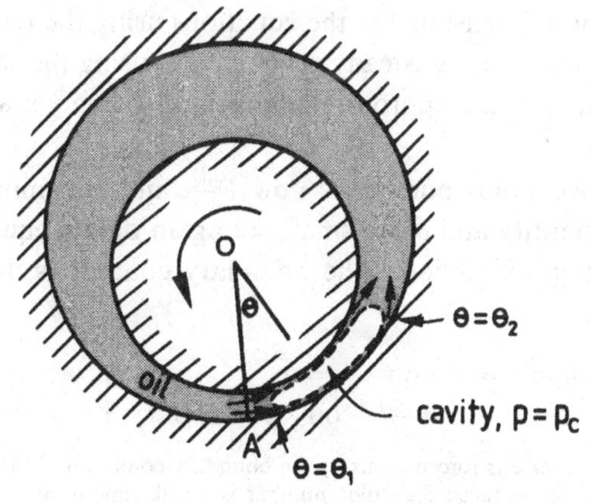

Fig. 4.7. Cavitation in a journal bearing

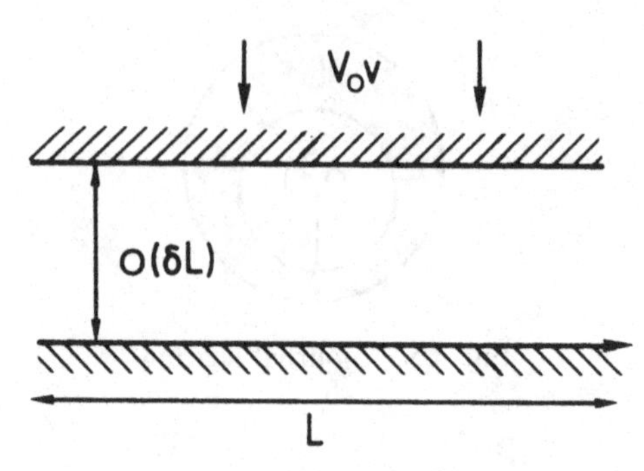

Fig. 4.8. A squeeze film

the three-dimensional case where the upper bearing surface is given in nondimensional terms by $z = \delta H(x, y)$ and the lower surface is moved with constant tangential velocity in the x-direction. As shown in exercise 8, this results in the equation

$$\nabla(\frac{1}{6}H^3\nabla p) = \frac{\partial H}{\partial x} \tag{4.13}$$

where

$$\nabla = \left(\frac{\partial}{\partial x}, \frac{\partial}{\partial y}\right).$$

4.2 Squeeze films

High pressures can also be generated in thin films by relative *normal* velocities across a thin film of fluid. In the situation shown in figure 4.8, we need to nondimensionalise the equations using the typical normal velocity V_0; this is now an unsteady problem and using the same scalings for x and y as in §4.1, the appropriate tangential velocity scale is $\frac{V_0}{\delta}$ and the time scale[2] is $\frac{\delta L}{V_0}$.

The reduced Reynolds number is now $\frac{V_0 \delta L}{\nu}$ and, assuming as before that both this quantity and δ are small, we again obtain equations (4.5)-(4.7) as a first approximation. The boundary conditions, however, are changed to

$$\left.\begin{array}{lll} u = v = 0 & \text{on} & y = 0 \\ u = 0, \quad v = \frac{\partial H}{\partial t} & \text{on} & y = H(x,t) \end{array}\right\}. \tag{4.14}$$

and

[2] This choice of time scale is forced on us by the boundary condition (4.14) on the moving plate. As long as the reduced Reynolds number is small, time derivatives can still be neglected in the momentum equation (4.5).

Solving equations (4.5) and (4.6) leads to

$$u = \frac{1}{2}\frac{\partial p}{\partial x}y(y-H)$$

and then, from (4.7) and (4.14)

$$\frac{\partial H}{\partial t} = \int_0^H \frac{\partial v}{\partial y}dy = -\int_0^H \frac{\partial u}{\partial x}dy = -\frac{\partial}{\partial x}\int_0^H u dy = \frac{\partial}{\partial x}\left(\frac{H^3}{12}\frac{\partial p}{\partial x}\right). \tag{4.15}$$

In the simplest case of a flat plate when $H = H(t)$, it is straightforward to calculate the load on the squeeze film. It can further be shown that it will take an infinite time for the plates to make contact if the force on the upper plate is finite (exercise 3)! Of course this prediction is made by a model that neglects surface roughness;[3] including the compressibility of the fluid or any plate curvature which may be present does not eliminate this phenomenon, whereas porosity in the plates does.

4.3 Thin films with free surfaces

4.3.1 Zero surface tension

Another interesting application of the lubrication model is to the flow of a thin film with a free boundary. Such a film might move under gravity as, for example, the spread of molten lava in a volcanic eruption or the motion of a raindrop down a windowpane.

We begin by considering gravity-driven flow on a horizontal surface (figure 4.9); we can work out the correct scalings by seeing that, with the usual nondimensionalisation, equation (4.6) with a gravity term added is

$$0 = -\frac{\partial p}{\partial y} - \frac{\delta^3 g L^2}{U\nu}. \tag{4.16}$$

Since gravity is the driving force we define U as $\frac{\delta^3 g L^2}{\nu}$ where δ, L are the initial depth and length of the film.

At the free surface we now have the kinematic boundary condition

$$v = \frac{\partial H}{\partial t} + u\frac{\partial H}{\partial x} \quad \text{on} \quad y = H(x,t) \tag{4.17}$$

and we need two more conditions since the position of this boundary is unknown. These conditions come from the fact that if surface tension is neglected, there is no stress applied on the free surface. From (4.11) this

[3] See, for example, Bush, pp. 132–40.

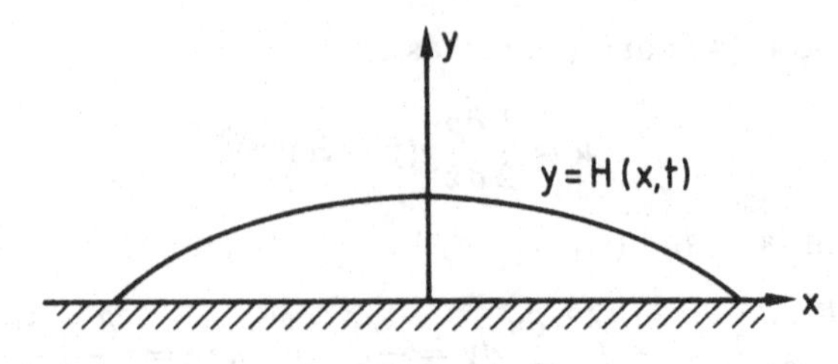

Fig. 4.9. A thin film

condition leads to

$$p = 0 \tag{4.18}$$

and

$$\frac{\partial u}{\partial y} = 0 \quad \text{on} \quad y = H(x,t). \tag{4.19}$$

Now, solving (4.16) and using (4.18), we get

$$p = -y + H$$

and then integrating (4.5) and using the stress-free boundary condition (4.19) gives

$$u = -\frac{1}{2}\frac{\partial H}{\partial x} y(2H - y).$$

Finally, using (4.7) and the kinematic boundary condition (4.17) leads to

$$\frac{\partial H}{\partial t} = \frac{\partial}{\partial x}\left(\frac{1}{3}H^3\frac{\partial H}{\partial x}\right). \tag{4.20}$$

This is a version of the *porous medium equation* for $H(x,t)$ which is a nonlinear version of the famous parabolic heat conduction equation. It is possible to find solutions of (4.20) which have 'compact support' (i.e., $H \equiv 0$ for all sufficiently large values of $|x|$ (exercise 5)) whereas it is not possible to find such solutions for the usual heat conduction equation $\frac{\partial H}{\partial t} = \frac{\partial^2 H}{\partial x^2}$.

A similar analysis is possible when the film lies on an inclined, or even a vertical surface. For a surface inclined at a fixed angle α, the force due to gravity comes into the x-momentum equation (4.5) and we have to choose U to make $\frac{h^2 g \sin\alpha}{U\nu}$ of $O(1)$. Then (4.6) is unchanged as long as $\alpha \gg \frac{h}{L}$ and we end up with a first order *hyperbolic* partial differential equation for $H(x,t)$ (see exercise 4). The fact that we get a hyperbolic equation in this case shows that discontinuities or 'shocks' can develop in the profile $H(x,t)$ as t increases.

4.3.2 Flows driven by surface tension

Surface tension is a very important mechanism for small scale flows such as paint films, the motion of a contact lens on the eyeball or various wetting or coating flows. Now, in two dimensions, the free boundary condition is $\sigma_{ns} = 0$, $\sigma_{nn} = T/R$ where s and n are tangential and normal coordinates respectively, T is the surface tension and R the radius of curvature of the surface of the film which is approximately $L/(\delta\partial^2 H/\partial x^2)$. Thus, with a suitable scaling for the pressure, (4.18) is replaced by $p = -\partial^2 H/\partial x^2$ and hence the thickness of a film on a flat base that is flowing under the influence of surface tension and viscosity can be shown (exercise 6) to satisfy a fourth-order evolution equation (sometimes called the Landau Levich equation)

$$\frac{\partial H}{\partial t} = -\frac{\partial}{\partial x}\left(\frac{1}{3}H^3\frac{\partial^3 H}{\partial x^3}\right). \tag{4.21}$$

Like (4.20), this is an equation that allows solutions with compact support.

Equations such as (4.21) can form the bases of models for several of the situations mentioned above. For example, in modelling the tear film in the vicinity of a circular contact lens moving in the x-direction on a flat eyeball, we could use the two-dimensional form of (4.21) with the addition of a convective term $\frac{\partial H}{\partial x}$ (exercise 8). However, to model paint films or foams, it may be important to take surface tension *gradients* into account, giving rise to a different extra term in equation (4.21) (exercise 7). Such gradients give rise to what are called *Marangoni flows*, and they have unexpectedly been found to dominate many zero-gravity fluid dynamics experiments carried out in space, in particular those concerned with crystal growth. The ability of the surface tension to vary spatially is also a crucial ingredient for the fluid to be able to form a foam (pure water has far too high a surface tension for foams to have any chance of surviving). It is also believed to be the mechanism responsible for the ripples that are often observed on solvent-based paint films.

4.4 Hele-Shaw flow

We will now abandon our "lubrication" motivation and consider the special case of a *Hele-Shaw cell* which involves the slow flow of a fluid between two *parallel* flat plates which are *fixed* at a small distance h apart. In this case $H \equiv 1$ and the reference velocity U is determined by the fluid flow which is now generated by some external pumping mechanism. The arguments leading to equations (4.5)-(4.7) can be used in three dimensions and lead, in the absence of gravity, to the thin film equations (really just "Couette flow" in three-dimensions)

$$\left.\begin{aligned} 0 = -\frac{\partial p}{\partial x} + \frac{\partial^2 u}{\partial z^2}, \quad 0 = -\frac{\partial p}{\partial y} + \frac{\partial^2 v}{\partial z^2}, \\ 0 = \frac{\partial p}{\partial z}, \quad \frac{\partial u}{\partial x} + \frac{\partial v}{\partial y} + \frac{\partial w}{\partial z} = 0. \end{aligned}\right\} \tag{4.22}$$

The boundary conditions are $u = v = w = 0$ on $z = 0, 1$. Hence we obtain

$$u = -\frac{1}{2}\frac{\partial p}{\partial x} z(1-z)$$

$$v = -\frac{1}{2}\frac{\partial p}{\partial y} z(1-z)$$

and integrating the continuity equation from $z = 0$ to 1 gives

$$-\frac{1}{12}\left(\frac{\partial^2 p}{\partial x^2} + \frac{\partial^2 p}{\partial y^2}\right) = 0 \tag{4.23}$$

where p is a function of x, y alone[4]. We also note that the *mean velocity* in the plane of the Hele-Shaw cell is

$$(\bar{u}, \bar{v}) = \int_0^1 (u, v)dz = -\frac{1}{12}\nabla p. \tag{4.24}$$

Thus equations (4.23) and (4.24) show that for this problem there is a "mean velocity" potential $-\frac{1}{12}p$, which satisfies Laplace's equation.

We now consider two classes of Hele-Shaw flow:

1. **Flow past an obstacle.** When a cylindrical obstacle is placed in the gap in a Hele-Shaw cell (figure 4.10), (4.22) and (4.23) lead us to expect that, because it is the gradient of a potential function, the mean velocity field for uniform flow past the obstacle will be related to that for *inviscid* irrotational flow in two dimensions. There, the fluid velocity $\mathbf{u} = \nabla\phi$ with $\nabla^2\phi = 0$, $\frac{\partial \phi}{\partial n} = 0$ on the obstacle and $\phi \to x$ at infinity. Here we have that the mean flow $\bar{\mathbf{u}} = -\frac{1}{12}\nabla p$ where $\nabla^2 p = 0$ away from the boundaries. The boundary conditions are $\bar{\mathbf{u}} \to (1, 0)$ at the edge of the cell and, if the obstacle is small compared with the length of the cell, this may be

[4] Note that this equation could be derived from (4.13) by putting $H = 1$. However this would be dangerous because of the different nondimensionalization used in the two cases.

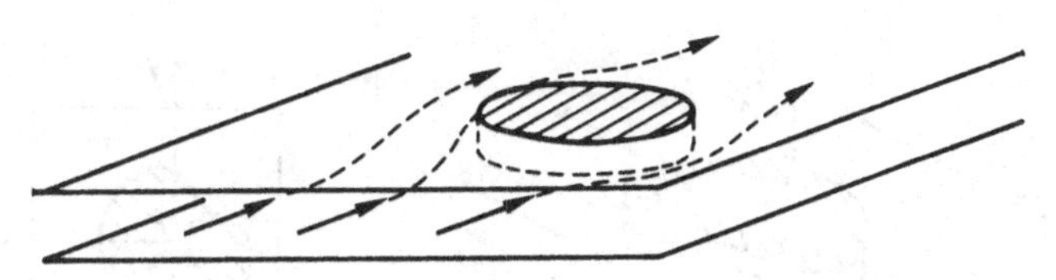

Fig. 4.10. Hele-Shaw flow past an obstacle

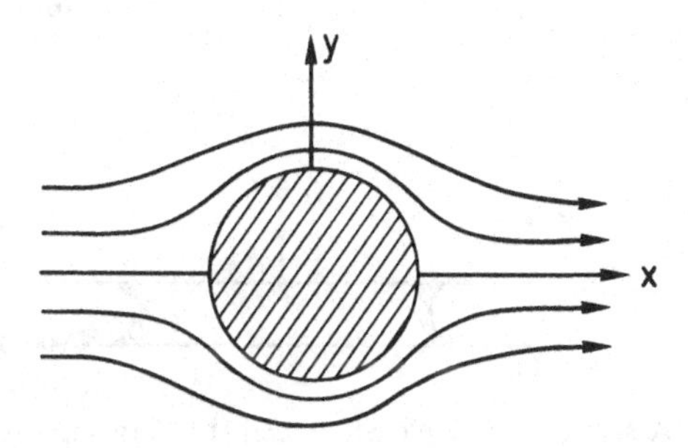

Fig. 4.11. Streamlines in Hele-Shaw flow past a circular cylinder

taken as $p \sim -12x$ at infinity. For the Hele-Shaw problem, we also need to take the normal velocity to be zero on the cylinder so that $\frac{\partial p}{\partial n} = 0$ there, and now, identifying ϕ with $-\frac{1}{12}p$, the two problems are identical. However, for the Hele-Shaw problem, the boundary condition on the obstacle needs further consideration since in this *viscous* flow problem we expect the tangential component of velocity to be zero as well as the normal component. However, we already have a well posed problem for p and we are not able to impose this extra tangential boundary condition. The situation is that the full noslip condition can only be achieved by introducing a 'boundary layer' of thickness $O(h)$ near the obstacle in the (x, y) plane in which the *full three-dimensional* slow flow equations (3.2) apply. The condition that the outer flow should match with this boundary layer is merely $\frac{\partial p}{\partial n} = 0$ and hence we expect that the model (4.23) is accurate away from this layer. Experimental observations in Hele-Shaw cells can readily be made and the streamlines can be seen to agree with potential flow predictions very closely. For streaming flow past a circular cylinder, the potential flow given by $p = -12(r + \frac{1}{r})\cos\theta$, which predicts streamlines symmetric fore and aft of the obstacle, can be clearly seen to occur in practice (figure 4.11 and Van Dyke (2), p. 8). Thus the *viscous dominated* Hele-Shaw flow faithfully reproduces the *zero* viscosity Euler model! In this case the 'boundary layer' mentioned above does *not* cause separation and remains thin over the whole of the boundary

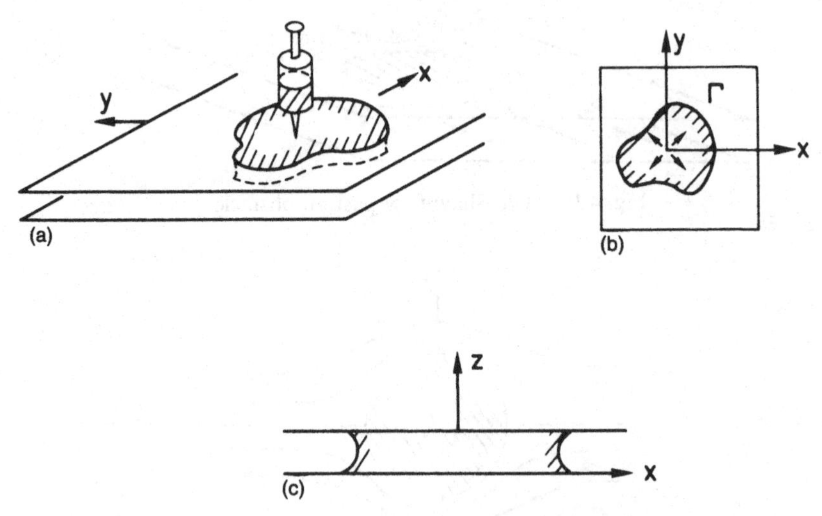

Fig. 4.12. A source in a Hele-Shaw cell (b) Plan view (c) Side view

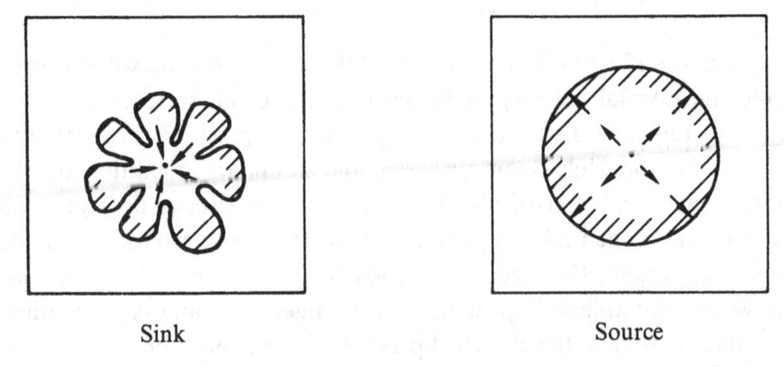

Fig. 4.13. Comparison of a source and a sink in Hele-Shaw flow

of the cylinder. An even more dramatic example of this phenomenon occurs when a cylinder with aerofoil cross-section is introduced into the cell. Then, even at high angles of incidence, the observed flow is the zero viscosity solution *without* circulation, as shown in Figure 2.12. The Kutta-Joukowski condition at the trailing edge does not apply here since it was a prediction of the Prandtl boundary layer theory for the aerofoil. It should be noted that although the streamlines are the same in these two situations, the pressure, which is identified with the velocity potential in the Hele-Shaw cell but is determined by Bernoulli's equation in inviscid flow, is quite different in the two cases.

2. **Flows with free boundaries in a Hele-Shaw cell.** A particularly interesting observation can be made when a Hele-Shaw cell is only partially filled with fluid as is the case when fluid is injected into or sucked from the cell at the origin, as shown in figure 4.12, thus creating a source or sink. The simplest assumption we can make about the conditions at the free surface Γ is that the pressure is constant. This assumes that the dominant *surface tension* force is constant, which means that the radius of curvature of the elevation of the advancing or retreating surface (shown in figure 4.12c) is constant. The other condition at the surface Γ given by $f(x, y, t) = 0$ is the kinematic condition averaged in the z-direction so the boundary conditions are

$$p = 0 \quad \text{and} \quad \frac{Df}{Dt} = 0 \quad \text{on} \quad f = 0 \tag{4.25}$$

where

$$\frac{D}{Dt} = \frac{\partial}{\partial t} + \bar{u}\frac{\partial}{\partial x} + \bar{v}\frac{\partial}{\partial y}.$$

These conditions can be formulated economically by using complex variable notation (see exercise 9). The dramatic observation here is that even though the model is reversible (replacing t by $-t$ and p by $-p$ leaves the equations unchanged) an *advancing* fluid interface due to a source is stable whereas a *retreating* interface due to a sink is unstable and breaks into 'fingers' (figure 4.13). We will consider this stability problem in more detail in the next chapter, but we remark here that the situation is reminiscent of the difference between melting ice (the phase interface becomes smoother and smoother as an ice cube from the refrigerator melts) and freezing it, when irregular "dendrites" of ice form (either around the edge of the ice tray, or in the formation of a snowflake). Indeed, when we identify p with the temperature, this model is a special case of what is called the *Stefan model* for freezing and solidification (exercise 10). Equation (4.25) states that phase change occurs at a prescribed temperature and that there is a release or uptake of latent heat there. A similar situation can also be observed in flow through a porous medium as described in the next chapter.

Exercises

1 A slider bearing consists of a flat plate $y = 0$ moving with constant velocity U in the x-direction and a fixed slider at $y = h(1 - \lambda\frac{x}{L})$ for $0 < x < L$ and $0 < \lambda < 1$. Given that $\frac{Uh^2}{\nu L} \ll 1$,

and the ambient pressure outside the bearing is p_0, show that the pressure within the bearing is

$$p_0 + \frac{6\mu UL\lambda x(L-x)}{(2-\lambda)h^2(L-\lambda x)^2}.$$

Show that the total load supported by the bearing is

$$\frac{6\mu UL^2}{\lambda^2 h^2}\left[\ln\frac{1}{1-\lambda} - \frac{2\lambda}{2-\lambda}\right].$$

2 A cylindrical shaft of radius a rotates with angular velocity ω within a fixed cylindrical housing (journal) of radius $a(1+\epsilon)$. The gap between the cylinders is filled with oil of viscosity ν. Show that when the centre of the shaft is offset a distance $d\epsilon$ ($d < a$) from the centre of the journal, the gap is given by $h(\theta) = \epsilon(a - d\cos\theta)$ when ϵ is small and θ is the polar angle shown in figure 4.5. Deduce that the lubrication approximation is valid in the gap if $\omega a^2\epsilon^2 \ll \nu$. Hence show that the pressure p satisfies

$$\frac{dp}{d\theta} = \frac{6(H-H_0)}{H^3}$$

where $H = 1 - \frac{d}{a}\cos\theta$ and H_0 is a constant given by

$$H_0 = \frac{\int_0^{2\pi} d\theta/H^2}{\int_0^{2\pi} d\theta/H^3}.$$

Deduce that $\int_0^{2\pi} pd\theta\cos\theta = 0$ and hence that the force on the shaft is perpendicular to the line of centres ($\theta = 0$). What are the implications for the trajectory of the shaft (assuming it is unrestrained at its ends and that the journal is fixed)?

3 A squeeze film exists between two parallel plates of length L at $y = 0$ and $y = \delta L$. Starting at $t = 0$, a constant force F is exerted downwards on the top plate. Show that the correct scaling for the horizontal velocity is $\frac{F\delta^2}{\mu}$. Deduce that at time t later the distance between the plates is $\frac{\delta L}{\sqrt{1+\alpha t}}$ where α is to be determined. This predicts that the plates will not touch in finite time. Consider the validity of this prediction by computing roughly the time taken for the film to close when the surface roughness (the "asperity" size) has amplitude $10^{-3}\delta L$.

4 A thin layer of rain water of viscosity ν is draining under the effects of gravity down the side of a vertical flat windowpane. If its thickness h is a function only of time t and distance x down the pane, show that the lubrication theory approximation leads to $h(x,t)$ satisfying the equation

$$\frac{\partial h}{\partial t} + \frac{gh^2}{\nu}\frac{\partial h}{\partial x} = 0.$$

Verify that $h = f\left(x - \frac{gh^2}{\nu}t\right)$ solves this equation, so that particular values of h propagate down with speed proportional to h^2. Hence, or otherwise, draw rough sketches of the evolution with time of a realistic initial profile $h(x,0)$. How would your results change if the plate were inclined at an angle α to the horizontal?

5 Show that a solution of (4.19) that satisfies the condition

$$H(x,0) = \begin{cases} (1 - \frac{9}{10}x^2)^{\frac{1}{3}} & \text{if } |x| < \sqrt{\frac{10}{9}} \\ 0 & \text{if } |x| > \sqrt{\frac{10}{9}} \end{cases}$$

is

$$H(x,t) = \begin{cases} (1+t)^{-\frac{1}{5}}\left(1 - \frac{9}{10}\frac{x^2}{(1+t)^{\frac{2}{5}}}\right)^{\frac{1}{3}} & \text{if } |x| < \sqrt{\frac{10}{9}}(1+t)^{\frac{1}{5}} \\ 0 & \text{if } |x| > \sqrt{\frac{10}{9}}(1+t)^{\frac{1}{5}} \end{cases}.$$

Sketch H as a function of x for several values of t and discuss the circumstances under which this type of solution might model the evolution of a volcano.

6 A thin film (e.g. paint) flows on a horizontal surface under the action of a constant surface tension T. Show that the appropriate scaling for the horizontal velocity is $\delta^3 T/\mu$ and that the boundary conditions on the surface $y = H(x,t)$ are

$$\frac{\partial u}{\partial y} = 0 \quad \text{and} \quad p = -\frac{\partial^2 H}{\partial x^2}.$$

Assuming that $g \ll \frac{T}{\rho L^2}$, solve the lubrication equations and show that H satisfies the equation

$$\frac{\partial H}{\partial t} = -\frac{1}{3}\frac{\partial}{\partial x}\left(H^3\frac{\partial^3 H}{\partial x^3}\right).$$

*7 If the surface tension of a fluid can vary, it will induce a shear stress in the free surface given by $\sigma_{ns} = \frac{\partial T}{\partial s}$. Show, by writing $T = T_0\sigma(x,t)$ and making the same assumptions as in exercise 6, that the nondimensional boundary conditions on $y = H$ are approximately

$$p = -\sigma\frac{\partial^2 H}{\partial x^2} \quad \text{and} \quad \delta^2\frac{\partial u}{\partial y} = \frac{\partial \sigma}{\partial x}.$$

Assuming that the surface tension varies slowly so that σ may be written as $1 + \delta^2\sigma_1(x,t)$, show that H will approximately satisfy the equation

$$\frac{\partial H}{\partial t} = -\frac{1}{3}\frac{\partial}{\partial x}(H^3\frac{\partial^3 H}{\partial x}) - \frac{1}{2}\frac{\partial}{\partial x}(H^2\frac{\partial \sigma_1}{\partial x}).$$

In a simple model for the drying of paint by heating from below, we can assume that $\sigma_1 = \frac{\lambda}{H}$ where $\lambda > 0$. The paint is applied by brush so that $H(x,0) = 1 + \varepsilon \sin kx$ where ε is a small parameter. Show that this model predicts that the amplitude of surface oscillations will decay and find the time scale for this decay.

*8 A slider bearing of length L consists of a flat plate $z = 0$ moving with constant velocity U in the x-direction and the slider which is given by $z = hH(x,y)$ where $H(x,y)$ is of $O(1)$. Explain carefully the conditions under which the lubrication approximation can be used and derive Reynolds' equation

$$\frac{\partial H}{\partial x} = \frac{1}{6}\left[\frac{\partial}{\partial x}\left(H^3\frac{\partial p}{\partial x}\right) + \frac{\partial}{\partial y}\left(H^3\frac{\partial p}{\partial y}\right)\right]$$

where p is the nondimensional pressure.

A model for a contact lens sliding over an eyeball is made by modelling the lens as a flat circular plate of radius a, a distance h from the plane $z = 0$ (the eyeball) that is moving steadily with velocity U in the x-direction relative to the lens. Tear fluid extends under the lens and covers the eyeball to a depth $hH(x,y)$. Assuming that the surface tension condition at the free surface is approximately $\sigma_{nn} = \frac{Th}{a^2}\nabla^2 H$, show that H satisfies the equation

$$\frac{\partial H}{\partial x} = \lambda\left[\frac{\partial}{\partial x}\left(H^3\frac{\partial}{\partial x}(\nabla^2 H)\right) + \frac{\partial}{\partial y}\left(H^3\frac{\partial}{\partial y}(\nabla^2 H)\right)\right]$$

for $x^2 + y^2 > 1$, where $\lambda = Th^3/3\mu Ua^3$. Justify the use of the lubrication approximation for this application and estimate λ given that $T = 10^{-1}$ N/m for tear fluid.

9 Derive the equations and boundary conditions for the flow of liquid in a Hele-Shaw cell with a free boundary. Show that, if the free boundary is given by $f(x, y, t) = 0$, the boundary condition may be written in terms of suitable nondimensional variables as

$$\frac{\partial p}{\partial t} - \frac{1}{12}(\nabla p)^2 = 0 \quad \text{on } f = 0.$$

Deduce that if $p = \text{Rl } w(z, t)$ where $z = x + iy$ then

$$\text{Rl} \frac{\partial w}{\partial t} = \frac{1}{12} \left| \frac{dw}{dz} \right|^2 \quad \text{on } f = 0.$$

The Hele-Shaw cell is driven by a constant sink of strength Q, so that $p \sim Q \log r$ as $r \to 0$, and $f = r - 1$ at $t = 0$. By writing $f = r - s(t)$ where s is the radius of the fluid disk, determine p and s. Show that the liquid will all be removed when $t = \frac{6}{Q}$.

*10 The temperature T in a stationary heat-conducting material with constant properties satisfies

$$\rho c \frac{\partial T}{\partial t} = k \nabla^2 T.$$

Suppose the material undergoes a phase change (e.g. melts or freezes) at $T = T_m = \text{const.}$ and that the latent heat is L per unit volume. Show that at the phase boundary with normal $\mathbf{n}$ there is a jump in heat flux $-k\nabla T$ such that

$$[-k\nabla T.\mathbf{n}]_{\text{solid}}^{\text{liquid}} = \rho L v_n$$

where v_n is the velocity of the phase boundary in the direction $\mathbf{n}$. Deduce that if this boundary is $f(x, y, z, t) = 0$,

$$[k\nabla T.\nabla f]_{\text{solid}}^{\text{liquid}} = \rho L \frac{\partial f}{\partial t}.$$

Show that this "Stefan" model reduces to the Hele-Shaw free boundary problem (4.25) when we (i) put $c = 0$, (ii) consider the solid to be at constant temperature $T_m - 0$ and (iii) set $T_{\text{liquid}} - T_m = p$. Note that $p > 0$ corresponds to $T_{\text{liquid}} - T_m > 0$, i.e., $p < 0$ corresponds to "supercooled" liquid.

*11 Show that the dimensionless Reynolds equation

$$\frac{d}{dx}\left(\frac{h^3}{6}\frac{dp}{dx}\right) - \frac{dh}{dx} = 0 \quad -0 < x < 1$$

is the Euler-Lagrange equation for the variational problem of minimising

$$\int_{-0}^{1}\left[\frac{h^3}{12}\left(\frac{du}{dx}\right)^2+\frac{dh}{dx}.u\right]dx \qquad (*)$$

over suitable u which satisfy the same end conditions as p.

Suppose now that the solution predicts that p can fall below the cavitation pressure, say $p = 0$. We could then model the ensuing cavity by saying that there is a boundary (unknown in advance) at which the solution of Reynolds' equation satisfies $p = 0$ and $\frac{dp}{dx} = 0$. Interpret this second boundary condition physically in terms of the flow pattern near the cavity.

Explain why it is reasonable that, even when cavities are present, the solution could be found by minimising (*) but restricting u to be *positive* in $(0, 1)$. (This is called a *variational inequality*.)

*12 Show the pressure p in a viscous liquid flowing in a thin annulus between fixed concentric spheres satisfies the "Laplace-Beltrami" equation

$$\frac{\partial}{\partial\theta}\left(\sin\theta\frac{\partial p}{\partial\theta}\right)+\frac{1}{\sin\theta}\frac{\partial^2 p}{\partial\phi^2}=0,$$

where θ, ϕ are spherical polar coordinates.

5

Spatial and temporal complexity

All the solutions considered so far have had very regular streamlines or particle paths but a glance at the weather or a smoke plume shows that spatial and temporal complexity arises in many common situations. In this chapter we will look at some situations where, in spite of this complexity, a mathematical model can be made of at least some aspects of the physical problem. We begin with the problem of flow in a porous medium where the spatial complexity can be *averaged* to give a smoothly varying macroscopic model.

5.1 Flow in a porous medium

Surprisingly, the Hele-Shaw model of Chapter 4 can be used as a model for the geometrically complicated problem of flow of a viscous fluid through a *porous medium.* Such a model is relevant in many practical situations such as oil recovery, hydrology, soil mechanics, filter design, and fluidized beds as well as numerous other phenomena in the earth sciences. The most basic model assumes that, besides there being an obvious *microscopic flow scale* defined by the 'pore' size and illustrated in figure 5.1, there is a much larger *macroscopic scale* over which the problem is to be studied; this macroscopic scale may be the overall dimensions of the industrial device or the oil field. We can make progress by working on an intermediate scale which is small compared to the macroscopic scale yet still contains enough pores for an *averaged* velocity $\mathbf{u}$ and pressure p to be defined. In figure 5.1 it can be seen that although the direction of the actual flow has large variations on the 'pore' size scale, the average velocity over a large number of pores will be a flow which goes from left to right. Both $\mathbf{u}$ and p will then be smoothly varying functions on the macroscopic scale. Although some progress has been

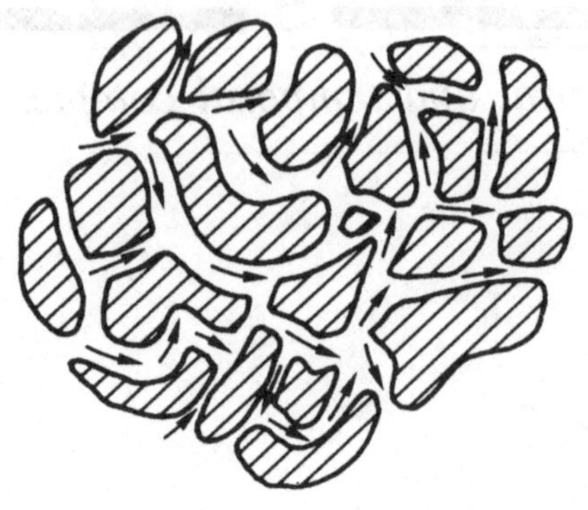

Fig. 5.1. Flow in a porous medium

made in averaging the slow flow equations over complicated geometries,[1] it is observed experimentally that averaged flow variables $\mathbf{u}$ and ∇p do exist which are constant on an appropriate intermediate scale and that, on the macroscopic scale, the flow through a porous medium in the absence of gravity obeys the law

$$\mathbf{u} = -\frac{k}{\mu}\nabla p \tag{5.1}$$

where k is the permeability and the quantity $\frac{k}{\mu}$ has the following properties:

1. It depends inversely on the viscosity μ.
2. It increases with 'porosity'[2] or 'void fraction' ϕ.
3. It decreases with the 'tortuosity' or 'fractal dimension' of the medium.

Different kinds of porous media are illustrated schematically in figure 5.2 and it can be seen that the above properties appear very plausible.

Equation (5.1) is known as *Darcy's law* and is a 'constitutive equation' for momentum transfer; it provides a much simpler relation between p and $\mathbf{u}$ than either (1.1) or (1.21). It is comparable to Fourier's law in heat transfer, Ohm's law in electromagnetism or Fick's law in diffusion or mass transfer. When any of these laws is coupled with the appropriate conservation equations, the field equations for the problem are obtained.

[1] See, for example, Ockendon and Ockendon.

[2] Caution: in many engineering applications (5.1) is assumed for the so-called average "superficial" velocity, i.e., the velocity which when distributed smoothly throughout the porous medium gives the same mass flow as does the true average velocity which exists just in the pore space. These two velocities are, of course, in the ratio of ϕ.

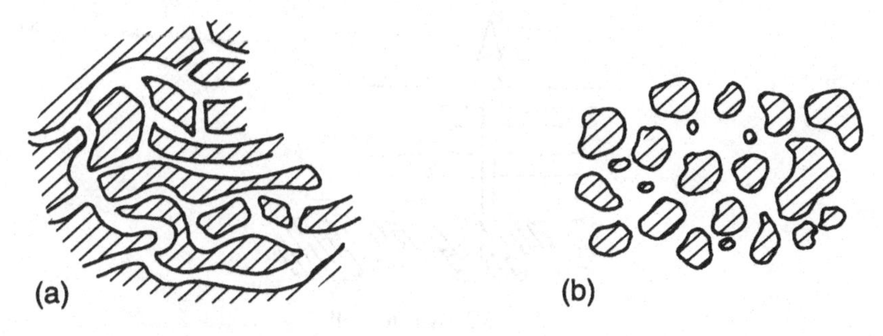

Fig. 5.2. Types of porous media (a) More tortuous, smaller ϕ (b) Less tortuous, larger ϕ

In this case we need only use conservation of mass in order to obtain a field equation for p. We note that Darcy's law may be seen as a three-dimensional generalization of equation (4.24) for a Hele-Shaw cell which, in dimensional coordinates, is

$$(\bar{u}, \bar{v}) = -\frac{h^2}{12\mu}\nabla p. \tag{5.2}$$

However, we emphasise that, while (5.2) applies to Hele-Shaw cells and hence to suitable porous media on the microscopic pore size when the reduced Reynolds number is small, the region of validity of the macroscopic law (5.1) is not so easy to determine theoretically.

If we assume that the fluid is incompressible then we can derive the equation of conservation of mass just as we did in Chapter 1. We note that the control volume can only extend over the fluid *in* the pores and so the continuity equation for the fluid is

$$\phi_t + \nabla.(\phi\mathbf{u}) = 0 \tag{5.3}$$

where ϕ, the void fraction, is the volume of pore space contained within a unit volume of the porous medium that contains fluid.[3] The equation for p, obtained from (5.1) and (5.3) is therefore

$$\phi_t = \nabla.\left(\frac{\phi k}{\mu}\nabla p\right). \tag{5.4}$$

In particular, for flow in a uniform medium we see that p is a harmonic function.

[3] $\phi\mathbf{u}$ is the 'superficial' velocity.

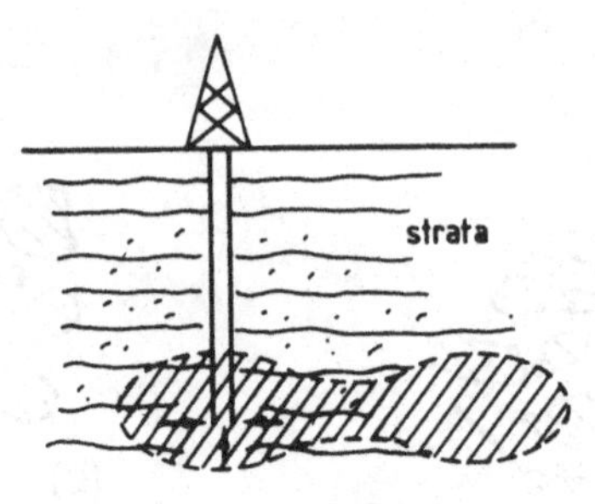

Fig. 5.3. An oil well

Experiment shows that the effect of gravity on these flows is just to add a hydrostatic component so that

$$\mathbf{u} = -\frac{k}{\mu}\nabla(p + \rho g z) \tag{5.5}$$

where z is measured vertically upwards.

Simple problems in hydrology or oil recovery can now be analysed (for more details see Yih). For example, we can model an oil well as a point sink and consider the flow of oil through the porous rock towards this sink (figure 5.3). At the sink, the pressure p will have a singularity analogous to the singularity of the velocity potential at a sink in irrotational inviscid flow (exercise 2) (but, as in chapter 4, we must beware of treating this analogy too naively).

One of the most interesting applications of this theory concerns flow where the porous medium is not completely saturated; that is the pores are not all full of the viscous fluid. In some circumstances (for example, wet soils) the pore space is filled with a mixture of liquid and vapour but the situation is easier to model when, as in an aquifer or a dam, there is a *region* of totally saturated pores abutting a region in which the pores are dry as shown in figure 5.4. In such problems there is a free boundary between the saturated and the unsaturated regions at which we take the mechanical boundary condition

$$p = p_0, \tag{5.6}$$

in the absence of capillary effects, and the kinematic condition is

$$\mathbf{u.n} = -\frac{k}{\mu}\frac{\partial}{\partial n}(p + \rho g z) = -\frac{\frac{\partial f}{\partial t}}{|\nabla f|} \tag{5.7}$$

on $f(x, y, z, t) = 0$. We note that these conditions are identical to those applied to a free boundary in a Hele-Shaw cell (4.25). We may therefore

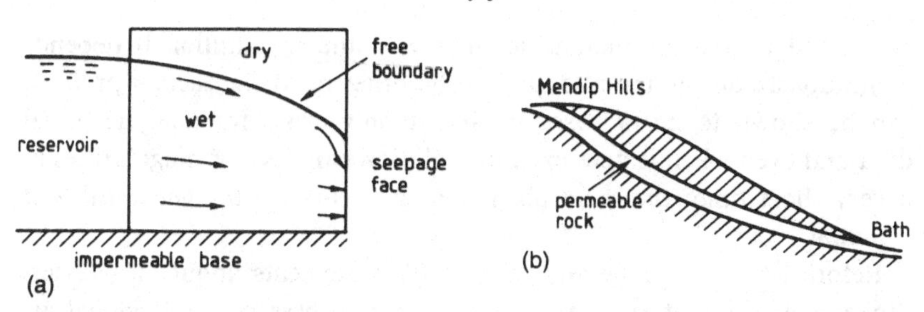

Fig. 5.4. Examples of semisaturated porous media (a) A semisaturated dam (b) An aquifer

expect a close similarity between these two problems. One example of this analogous behaviour occurs in the *extraction* of oil from a point sink which is notoriously unstable when the oil is surrounded by or being pumped by a less viscous fluid such as air or water; this instability (which results in 'coning') corresponds exactly to the instability of a Hele-Shaw cell described at the end of Chapter 4 and illustrated in figure 4.13.

5.2 Unsteady flows

Flow in a porous medium provides an example of *prescribed* spatial complexity; the *spontaneous* spatial or temporal evolution of complicated flow morphologies is much more difficult to describe mathematically. The principal obstacle to be overcome is the difficulty of solving the *unsteady* Navier-Stokes equations even approximately. We therefore review the few situations where we can make some 'global' remarks about evolution problems and then consider the implications of a 'local' stability analysis for some special problems. We will be following in the tradition of classical mechanics for rigid bodies where, although there are only a few cases in which Newton's equations can be solved exactly for all time, there are many cases in which a 'normal mode' analysis is possible for small displacements; such a solution will usually describe at least the initial evolution of the motion.

We begin by recalling that almost the only unsteady flows about which we have been able to say anything so far are the thin film problems in Chapter 4 and the porous medium free boundary problem in §5.1. In these cases the time derivative only entered via the kinematic boundary condition and not in the field equation. Even so difficulties can arise. As described in §4.4, the Hele-Shaw *injection* problem is *well posed*; that is to say, it is possible to prove that the solution to the initial value problem exists and is free of singularities in $t > 0$ and in addition it depends continuously on the initial data. On the other hand the *suction* problem can be shown to be *ill posed* in that it only exists for analytic initial data and even when it does exist the solution can develop singularities in a very short time; the same phenomenon occurs for the *backward heat equation.*

Before we try to make any comparable statements about the Navier-Stokes equations, it is instructive to see how these provable *global* assertions about Hele-Shaw or porous medium flow fit in with a *local* stability analysis of a particular flow. Suppose that we consider a basic one-dimensional flow in a uniform porous medium in which there is a free boundary at $x = Vt$, and gravity may be neglected. The fluid is in $x < Vt$ where the pressure is $-\frac{\mu V}{k}(x - Vt)$ and the velocity is $(V, 0, 0)$ thus satisfying (5.1), the continuity equation, and boundary condition (5.6). We now seek a solution in which a small perturbation or disturbance is imposed on this simple flow (figure 5.5). As in the theory of Stokes waves in irrotational flow, say, we proceed by writing down the problem for the perturbation pressure $\hat{p}$ where

$$p = -\frac{\mu V}{k}(x - Vt) + \hat{p}$$

and, for convenience, change to $x - Vt = \xi$ as a new space variable. We find, since p is a harmonic function, that

$$\frac{\partial^2 \hat{p}}{\partial \xi^2} + \frac{\partial^2 \hat{p}}{\partial y^2} = 0,$$

with $\hat{p} = \frac{\mu V \xi}{k}$ on the free boundary $\xi = h(y, t)$ say. When we consider the second free boundary condition (5.7) and neglect second-order terms in h and $\hat{p}$, we find

$$\frac{\partial \hat{p}}{\partial \xi} = -\frac{\mu}{k}\frac{\partial h}{\partial t}, \quad \hat{p} = \frac{\mu V h}{k} \quad \text{on} \quad \xi = 0.$$

Thus if we seek a perturbation[4] $h = h_0 e^{i\alpha(y-ct)}$ with wave number α and speed c and assume exponential decay as $\xi \to -\infty$, we find $\hat{p} = \frac{\mu V h_0}{k} e^{i\alpha(y-ct)+|\alpha|\xi}$, where $c = -iV|\alpha|/\alpha$ is the "dispersion relation" between the wave speed and wave number. The growth rate is $\mathrm{Rl}(-i\alpha c) = -V|\alpha|$ which suggests that the stability properties will depend dramatically on the sign of V. If $V > 0$, all perturbations except

[4] Note that the perturbation can always be decomposed into exponentials in space and time because the field equation and boundary conditions are linear and do not involve x, y or t explicitly.

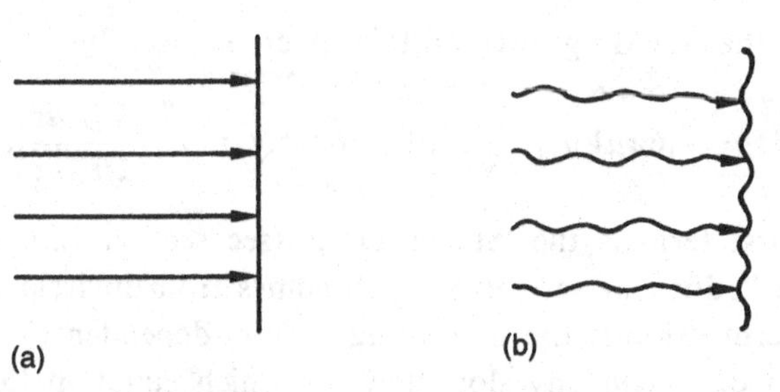

Fig. 5.5. Stability of flow in a porous medium (a) Streamlines for basic flow (b) Streamlines for perturbed flow

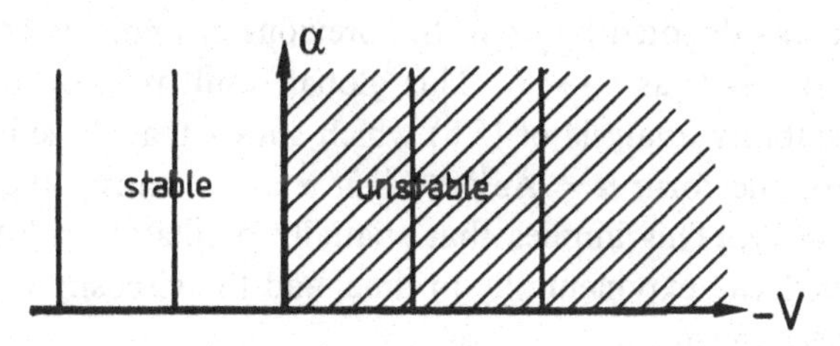

Fig. 5.6. The level curves of Imc

those with $\alpha = 0$ decay in time but if $V < 0$, so that the free boundary is retreating, all perturbations grow in time with the short wavelengths growing fastest (cf. figure 4.13). For future reference we plot the level curves of Imc on a $(\alpha, -V)$ plane in figure 5.6; the curve Imc $= 0$ separates the stable and unstable regimes.

We remark that it is possible to give a simple physical interpretation of this stability result because, if one of the protrusions in figure 5.5b were to grow out of the fluid region when $V > 0$, the pressure gradient in its vicinity would consequently decrease and thereby smooth the protrusion; the reverse is true when $V < 0$.

One other situation in which we can make progress with genuinely unsteady flow concerns the slow motion of rigid bodies in an unbounded viscous fluid. If the parameter regime is such that the slow flow model given in Chapter 3 is applicable and we consider a single sphere, the problem to be solved is

$$\frac{\partial \mathbf{u}}{\partial t} = -\frac{1}{\rho}\nabla p + \nu \nabla^2 \mathbf{u} \tag{5.8}$$

$$\nabla . \mathbf{u} = 0$$

with $\mathbf{u} = V(t)\mathbf{i}$ on the sphere $(x - \int_0^t V dt)^2 + y^2 + z^2 = a^2$. It can be shown (exercise 5) that the drag force on this sphere is given by

$$\mathbf{D} = -[6\pi\mu Va + \frac{2\pi}{3}a^3\rho\dot{V} + 6\rho a^2\sqrt{\nu\pi}\int_{-\infty}^{t}\frac{\dot{V}(\tau)d\tau}{\sqrt{t-\tau}}]\mathbf{i}. \qquad (5.9)$$

Here the first term is the 'Stokes Drag' (see section 3.3), the second term is the "added mass" term which comes from inviscid theory and the third term exhibits the interesting 'history-dependence' of the drag force which occurs in any slow flow for which equation (5.8) applies. From (5.9) it can be seen that if the sphere has mass M and is in free motion with $\mathbf{D} = M\dot{V}\mathbf{i}$ the velocity V can be proportional to $e^{-i\omega t}$ only if ω, which was denoted by αc in the previous example, is complex and hence that $|V| \to 0$ as $t \to \infty$. This global result is also obvious from the 'local' stability analysis of (5.8) which shows that there can only be a solution of the form $\mathbf{u} = \mathbf{A}e^{i(\alpha.x-\alpha ct)}$ if α and c satisfy the dispersion relation $ic = \nu\alpha$. This implies that spatially oscillatory solutions of *all* wavelengths decay exponentially in time, and the viscosity is acting as a damping mechanism.

As a contrast, we note that the local dispersion relation obtained in inviscid flow theory for Stokes waves[5] on water of depth h is $\alpha c^2 = g\tanh\alpha h$ which gives neutral stability and permits oscillations to persist in both space and time.

With the above ideas in mind, we will now discuss the rudiments of *linear stability theory* for the Navier-Stokes equations.

5.3 Stability theory

When the nonlinear inertia terms in the Navier-Stokes equations are important then analytical progress can only be made if (1) the unsteadiness is a small perturbation about a basic steady flow and (2), this perturbation is localised in space and time as was the case in the Hele-Shaw stability analysis. We consider the simplest two-dimensional case and write the Navier-Stokes equations as in (2.10):

$$\frac{\partial\omega}{\partial t} + \frac{\partial(\psi,\omega)}{\partial(y,x)} = \frac{1}{Re}\nabla^2\omega \qquad (5.10)$$

where ψ is the stream function and the vorticity $\omega = -\nabla^2\psi$. Exploiting our "localisation" assumption, we suppose that the fluid is unbounded in all directions and that the basic steady flow is unidirectional with

[5] See, for example, Acheson, Ch. 3.

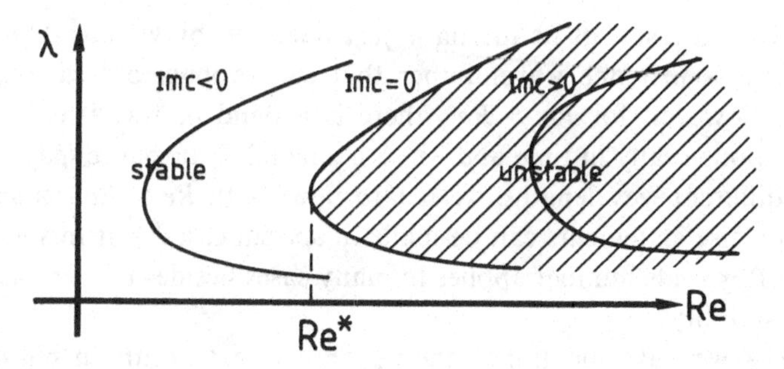

Fig. 5.7. The level curves of Imc for the Orr-Somerfeld equation

$\psi = \psi_0(y)$. In order to satisfy (5.10) we must have

$$\psi_0''''(y) = 0$$

and we seek an unsteady perturbation of the form

$$\psi \sim \psi_0(y) + \epsilon\psi_1(x, y, t) + \ldots$$

where ϵ is a small positive parameter measuring the size of the disturbance. Since the equation for ψ_1 will be linear with constant coefficients for the x and t derivatives, we again expect an exponential dependence on x and t and write

$$\psi_1 = g(y)e^{i\alpha(x-ct)}.$$

Substituting in (5.10) and equating coefficients of ϵ leads to the *Orr-Somerfeld* equation

$$\frac{1}{Re}(g'''' - 2\alpha^2 g'' + \alpha^4 g) = -i\alpha\psi_0''' g + i\alpha(\psi_0' - c)(g'' - \alpha^2 g). \qquad (5.11)$$

We have assumed that the disturbance is localised in the y direction so that $|g| \to 0$ as $|y| \to \infty$, and we study a wavelike variation in the x-direction, by taking α to be real and positive. Then, for any given α, (5.11) is an eigenvalue problem for c which can only be solved numerically. The general shape of the all-important contours Imc = constant are shown schematically in figure 5.7 for a particular choice of the cubic polynomial $\psi_0(y)$ (see Rosenhead, Ch. 9 for more details).

As in figure 5.6, the contour $Imc = 0$ divides the stable solutions with $Imc < 0$ from the unstable ones with $Imc > 0$. From figure 5.7 we can see that if the Reynolds number is small enough ($Re < Re^*$) then all disturbances are stable and decay in time. For $Re = Re^*$ we have the critical case where the tendency of the shear flow to "turn over" on itself as a result of its inertia is just balanced by viscous "diffusion" and there is just one wavenumber that can propagate at a real wave speed. However, for $Re > Re^*$, there is a band of wavelengths which grow exponentially in time and, since a general disturbance may contain many different wavelengths, we regard flows with $Re > Re^*$ as *unstable.* More precise statements can be made in special cases[6] but this idea of a *critical* Reynolds number applies to many cases besides this simple 'shear flow instability'.

In §3.4 we have mentioned the instability that occurs in blunt body flow when the wake changes from being steady to oscillatory as the Reynolds number increases. Other examples of instabilities are:

1. **The Taylor instability** occurs in fluid which is confined in an annulus between a fixed outer and a rotating inner circular cylinder. In the critical case the centrifugal forces just balance the viscous forces.

2. **The Rayleigh-Benard instability** which occurs in a fluid which is heated from below. The critical case occurs when the buoyancy of the fluid is just balanced by gravity. This is readily observable in the kitchen.

3. We have already remarked in §2.2 that the irreversibility of high Reynolds number jet flows explains why it is easier to blow out a candle than to suck one out (figure 5.8). The **Kelvin-Helmholtz** instability of the strong shear layer at the edge of a jet also implies that the jet evolves into a complicated, less unidirectional but still irreversible flow at large distances from the nozzle. The pressure variations in such a "turbulent" jet can be so extreme as to cause unpleasant acoustic effects.

We note that even though $\text{Im } c < 0$, there may be degeneracy in the eigenvalues of equations such as (5.11) which means that the solutions behave temporally like $t^k e^{-i\alpha ct}$ for some constant k. If k is positive this opens up the possibility that the disturbance could grow outside the validity of linear stability theory before the exponential decay has had a chance to set in.

5.4 Turbulence

Graphs like those in figures 5.6 and 5.7 suggest the following '*Landau*' theory for the transition from steady laminar flow to "turbulence". For

[6] For instance, it can be shown that when $Re = \infty$, so that $\psi_0'''' = 0$ no longer holds, the existence of an 'inflection point' in the flow (where $\psi_0''' = 0$) is necessary for instability to occur (exercise 6).

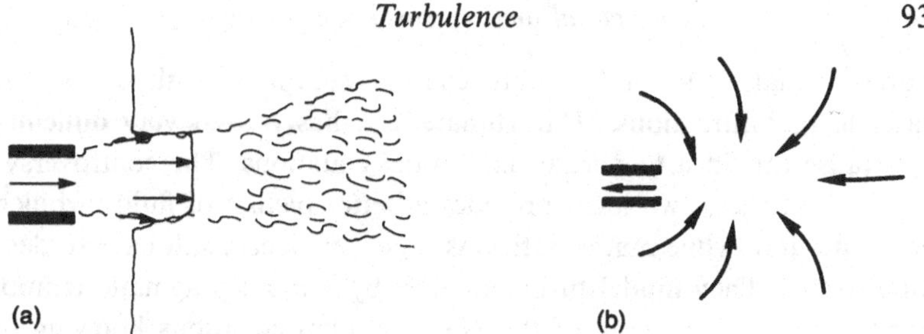

Fig. 5.8. Blowing and suction from an orifice (a) Blowing (b) Suction

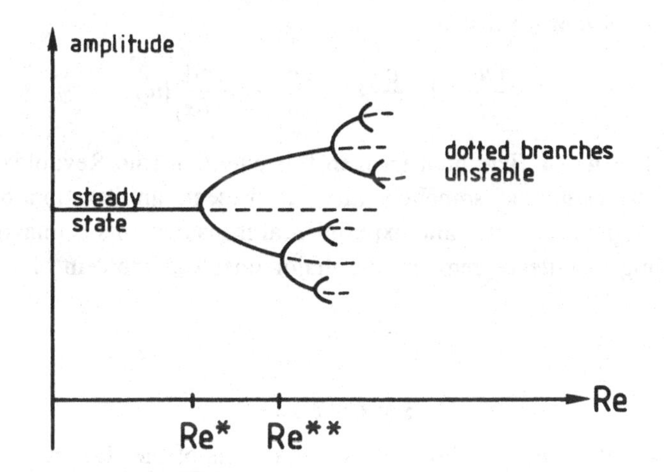

Fig. 5.9. Schematic response diagram for Landau theory

small values of some driving parameter (e.g., Re) there is a unique steady flow which is stable. At a critical value Re^* of the parameter, one or more wavelengths can support travelling waves and beyond this critical value there are some wavelengths at which disturbances grow temporally, and the steady flow is considered unstable. However, it can sometimes be shown that, for $Re > Re^*$, a new stable, possibly unsteady, solution has 'branched' from the steady flow and it may now remain stable for $Re^* < Re \leq Re^{**}$ say. We could then postulate that this process repeats itself indefinitely as shown diagramatically in figure 5.9. As the Reynolds number increases, more and more unsteady solutions come into existence, although many of them are unstable, and after many such 'bifurcations' we arrive at "*turbulence*".

A rival speculation emerged in the 1960s when it was found that a simple three-dimensional system of ordinary differential equations (the

Lorenz equations) could exhibit 'chaotic' solutions[7] without any such cascade of bifurcations. Unfortunately, it has proved very difficult to generalise this idea to partial differential equations. The controversy as to which of these two ideas provides a better picture of fluid turbulence continues but, while mathematicians argue, engineers still have to design aircraft etc. They model turbulent flow by a more pragmatic temporal and/or spatial averaging of the Navier-Stokes equations knowing that high frequency oscillations are expected in turbulent flow. The difficulty with such averaging is that the nonlinear terms do not 'average out' and so the *mean flow* $\bar{\mathbf{u}}$ satisfies

$$\rho\frac{\partial \bar{u}_i}{\partial t} + \bar{u}_j\frac{\partial}{\partial x_j}\bar{u}_i = \frac{\partial \bar{\sigma}_{ij}}{\partial x_j} - u'_j\frac{\partial}{\partial x_j}(u'_i)$$

where $\mathbf{u}' = \mathbf{u} - \bar{\mathbf{u}}$. The final term in this equation (the Reynolds stress) has to be estimated somehow and, at present, an iteration between the theoretical estimates and experimental measurements is unavoidable. Modelling turbulence remains the major unsolved problem of fluid dynamics.

5.5 Conclusions

In these notes we have discussed some aspects of the classical theory of viscous flows that are both physically relevant and susceptible to mathematical analysis. However, the modelling techniques we have employed enable us to describe many other continua besides. In particular, by relaxing our assumption that the stress tensor σ_{ij} is linearly related to the rate of strain tensor e_{ij}, we can consider a wide variety of materials as shown in the figure 5.10.

It is convenient to write $e_{ij} = \dot{\epsilon}_{ij}$ where $\dot{} = \frac{\partial}{\partial t} + \mathbf{u}.\nabla$ so that we can consider general linear constitutive relations between $\sigma_{ij}, \dot{\sigma}_{ij}, \epsilon_{ij}$ and $\dot{\epsilon}_{ij}$ and chart out many different partial differential equation models. The model which causes most debate is that of (mathematical) plasticity, a plastic here being defined as a *solid* which can *flow irreversibly* when some measure of the stress σ_{ij} attains a *yield criterion* (as in bending a paper clip); this means that models for plasticity are inevitably nonlinear. It is a great mathematical unification that the models can all be made to "merge into" each other, since viscous and elastic materials are both limits

[7] Chaotic solutions have the attribute that a small change in initial conditions can produce a dramatically different evolution pattern (which we certainly expect to be a characteristic of "turbulence").

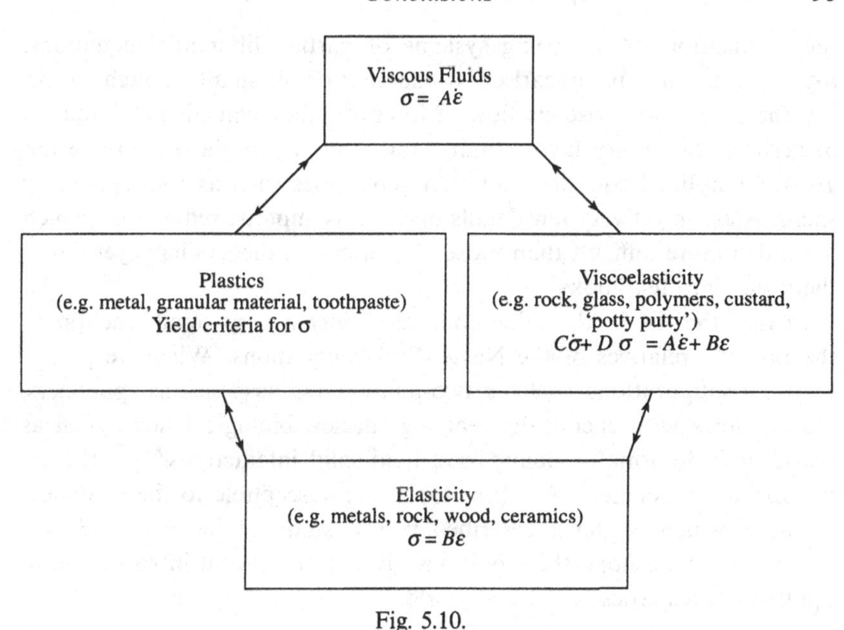

Fig. 5.10.

of viscoelastic materials and theories for viscoplastic and elastoplastic materials also exist.

Constructing such "composite" models requires considerable mathematical imagination. For example, the classical theory of elasticity almost always involves the compressibility of the material and there is no mention of any concept such as pressure. Hence a viscoelastic theory has to tend to a compressible, pressure-free model in one limit, and an incompressible model involving pressure and viscosity in another limit (in fact the "pressure" turns out to be a Lagrange multiplier in the incompressibility constraint when the elasticity model is formulated variationally). Similarly, a thermoelastic fluid is inevitably compressible, but in a different way from a gas. The usual way to model a gas is to postulate an equation of state between pressure, density, and temperature and close the model with the energy equation; however, since it is more natural to measure "coefficients of thermal expansion" in thermoelasticity, it is much more convenient to model this kind of material by assuming a temperature dependent stress-strain relation than it is to assume an equation of state.

As with the Navier-Stokes equations, all these models contain material parameters which can be allowed to vary in whatever manner may suit the application. Also, being systems of partial differential equations, they can usually be linearised if the motion is small enough, as in the theory of slow viscous flow. Moreover, they can all be simplified in certain "boundary layer" limits; for example, in elasticity there are greatly simplified equations for thin geometries such as rods, plates or shells. Alas, as yet very few details of these asymptotic reductions (which are much more difficult than those of lubrication theory) have yet found their way into textbooks!

In fact the list in the table comprises only a very small fraction of the possible relatives of the Navier-Stokes equations. When we permit hybrid configurations such as two-phase flow, suspensions, polymers, foams, flows with chemically reacting species, biological flows such as blood with deformable boundaries, fluid-solid interactions,[8] ..., the list is more or less endless. All these topics are susceptible to the analytical methods which we have described in our study of the Navier-Stokes equations, so we hope this book has given you a useful introduction to applied mathematics in the real world.

Exercises

1 Derive from first principles the continuity equation (5.3) for the flow of a fluid through a porous medium of void fraction $\phi(\mathbf{x}, t)$. Assuming that Darcy's law holds for the flow in the same medium under gravity, show that if both ϕ and the permeability are constant then

$$\nabla^2 p = 0.$$

2 An oil well is modelled as a point sink in a uniform porous medium from which oil is removed at a rate $q(t)$. If the oil field is regarded as infinite, show that the pressure p is given by

$$p = -\frac{q\mu}{4\pi k r} + p_\infty$$

where r is a spherical polar coordinate. Compare this with the pressure generated by a similar point sink in an inviscid incompressible fluid. (See figure 5.11.)

Water is extracted at a rate q from a well beneath the water table Γ. The boundary between dry and saturated soil, Γ, is

[8] A neat example is that of distinguishing between hard-boiled and fresh eggs. If you try to spin a sample you will easily see that the theory of precession of tops only applies to rigid bodies!

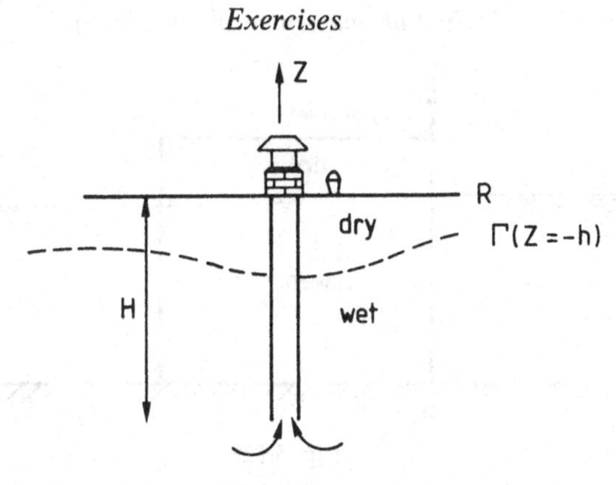

Fig. 5.11.

given by $z = -h(R)$, where $R^2 = x^2 + y^2$ and it is assumed that h is small compared with the depth of the well H (i.e. $q \ll \frac{k\rho g}{\mu}$). Write $\phi = -\frac{k}{\mu}(p + \rho g z)$ and justify the following model:

$$\nabla^2 \phi = 0 \quad \text{for} \quad z < -h(R)$$

with

$$\phi \sim \frac{q}{4\pi(x^2 + y^2 + (z + H)^2)^{\frac{1}{2}}} \quad \text{near} \quad (0, 0, -H)$$

and

$$\frac{\partial \phi}{\partial n} = 0, \quad \phi = +\frac{k\rho g h}{\mu} \quad \text{on} \quad z = -h.$$

Deduce that $h \simeq \frac{q\mu}{2\pi k\rho g(x^2+y^2+H^2)^{\frac{1}{2}}}$ and hence show that if the sink were a source the surface would *rise* by the same amount. Explain why this change in surface behaviour would *not* occur if a source or sink were placed under a free surface in inviscid flow.

3 The mean flow-velocity $\mathbf{u}$ in a saturated porous medium is given by Darcy's law

$$\mathbf{u} = -\frac{k}{\mu}\nabla(p + \rho g z)$$

in the usual notation. Explain the assumptions which permit the boundary conditions

(i) $\frac{\partial}{\partial n}(p + \rho g z) = 0$ at an impermeable fixed boundary, and

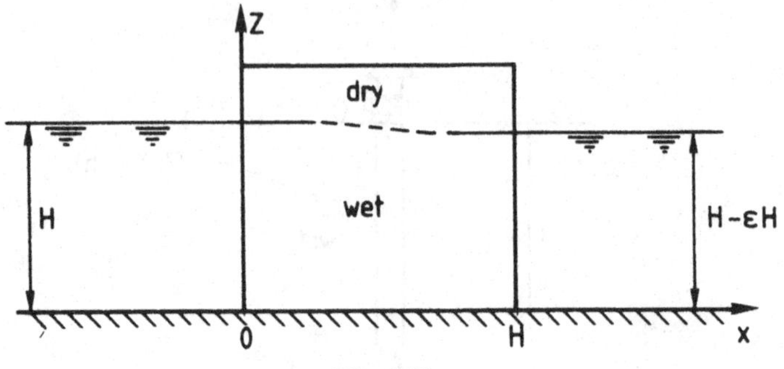

Fig. 5.12.

(ii) $p = 0 = \frac{\partial f}{\partial t} + (\mathbf{u}.\nabla)f$ at a boundary $f(x, t) = 0$ which separates a saturated region from a dry one which is open to the atmosphere.

In a porous medium, water flows in a shallow layer of thickness $h(x, t)$ above an impermeable base $z = 0$. The porous medium above the layer is dry and open to the atmosphere. Justify the approximate model $\mathbf{u} = (u, w)$ where

$$u = -\frac{k}{\mu}\frac{\partial p}{\partial x}, \quad 0 = \frac{\partial p}{\partial z} + \rho g, \quad \frac{\partial u}{\partial x} + \frac{\partial w}{\partial z} = 0$$

with $w = 0$ on $z = 0$, and $p = 0 = w - \frac{\partial h}{\partial t} - u\frac{\partial h}{\partial x}$ on $z = h$. Deduce that

$$\frac{\mu}{k\rho g}.\frac{\partial h}{\partial t} = \frac{\partial}{\partial x}\left(h\frac{\partial h}{\partial x}\right).$$

4 A square porous dam separates two reservoirs of depth H and $H - \varepsilon H$. If the pressure within the dam is given by $p = \rho g(H - z) + \varepsilon p_1$, show that, to lowest order in ε,

$$\nabla^2 p_1 = 0$$

with

$$\begin{aligned} \frac{\partial p_1}{\partial z} &= 0 \quad \text{on} \quad z = 0 \quad \text{and} \quad H, \\ p_1 &= 0 \quad \text{on} \quad x = 0, \\ \text{and} \quad p_1 &= -\rho g H \quad \text{on} \quad x = H. \end{aligned}$$

Solve this problem and show that the dam will be saturated to a height $H - \varepsilon x$. (See figure 5.12.)

*5 A sphere of radius a moves with velocity $U_0 V\left(\frac{t'}{t_0}\right)$ in a straight line through a viscous fluid where t' is dimensional time. Show that the Navier-Stokes equations can be nondimensionalized to obtain

$$\lambda \frac{\partial \mathbf{u}}{\partial t} = -\nabla p + \nabla^2 \mathbf{u}$$

where $\lambda = \frac{a^2}{\nu t_0}$ provided $\frac{U_0 a}{\nu} \ll 1$ and write down appropriate boundary conditions. By writing $\mathbf{u} = \operatorname{curl}(0, 0, \frac{f(r,t)\sin\theta}{r})$ show that f satisfies the equation

$$f_{rrrr} - \frac{4f_{rr}}{r^2} + \frac{8f_r}{r^3} - \frac{8f}{r^4} = \lambda(f_{rrt} - \frac{2}{r^2} f_t)$$

with $f(1,t) = \frac{1}{2}V(t)$, $f_r(1,t) = V(t)$ and $f \to 0$ as $r \to \infty$. By taking a Laplace Transform in t show that $\bar{f} = \int_0^\infty f e^{-pt} dt = \bar{V} g(r)$ where

$$g'''' - \left(\frac{4}{r^2} + \lambda p\right) g'' + \frac{8}{r^3} g' + \left(\frac{2\lambda p}{r^2} - \frac{8}{r^4}\right) g = 0.$$

Given that the solution of this equation which tends to zero as $r \to \infty$ is

$$g = \frac{A}{r} + B\left(\frac{1}{r} + \sqrt{\lambda p}\right) e^{-\sqrt{\lambda p} r}$$

show that

$$\bar{f} = \bar{V}(p)\left(\frac{1}{2\lambda p r}(3 + 3\sqrt{\lambda p} + \lambda p) - \frac{3}{2\lambda p}\left(\frac{1}{r} + \sqrt{\lambda p}\right) e^{\sqrt{\lambda p}(1-r)}\right).$$

The nondimensional drag on the sphere, given by (3.14), can be shown to be

$$\frac{4\pi}{3}\left[-f_{rrr} + \frac{2f_{rr}}{r} + \frac{2f_r}{r^2} - \frac{8f}{r^3} + \lambda f_{rt}\right]_{r=1}.$$

By evaluating the Laplace Transform of this quantity and then inverting, show that this drag is

$$-6\pi V(t) - \frac{2}{3}\pi\lambda \dot{V}(t) - 6\sqrt{\pi\lambda} \int_0^t \frac{\dot{V}(s)}{\sqrt{t-s}} ds$$

and hence that the dimensional drag is given by (5.9).

*6 An inviscid unidirectional flow between walls at $y = 0$ and 1 is given by a streamfunction $\psi_0(y)$. Consider a perturbation of the form $\psi = \psi_0(y) + \epsilon g(y) e^{ik(x-ct)}$ where $\epsilon \ll 1$ and show that

$$\int_0^1 \frac{\psi_0''' |g|^2}{(\psi_0' - c)} dy = - \int_0^1 (|g'|^2 + k^2 |g|^2) dy.$$

By considering the real and imaginary parts of this equation show that an instability is only possible if $\psi_0'''(y)$ changes sign in $0 < y < 1$.

[This result, due to Rayleigh, shows that the velocity must have a point of inflection for instability to occur.]

Appendix A

A brief introduction to asymptotics

This appendix is intended to give readers an intuitive idea of the theory of asymptotic expansions and in particular of what is meant by the symbols $\sim$, O and o, and the jargon "regular and singular asymptotic (or perturbation) expansions".

The starting point is the realisation that the familiar ideas associated with *convergent* series expansions are inadequate when one is trying to use truncated series to approximate functions which may, for example, be unknown solutions to differential equations or may be too complicated to comprehend without a computer at hand. Whereas it is often very convenient to approximate

$$\sin\varepsilon \simeq \varepsilon \quad \text{as} \quad \varepsilon \to 0, \tag{A.1}$$

what happens, say, with

$$\int_0^\infty \frac{e^{-t}}{1+\varepsilon t}dt \quad \text{as} \quad \varepsilon \to 0 \quad (\text{with } \varepsilon > 0 \text{ of course})? \tag{A.2}$$

The first case is easy to quantify because $\sin\varepsilon$ has a convergent Taylor series about $\varepsilon = 0$ and it can be proved that the error tends to $-\varepsilon^3/6$ as $\varepsilon \to 0$. We say

$$\sin\varepsilon \sim \varepsilon + O(\varepsilon^3) \quad \text{as} \quad \varepsilon \to 0$$

to mean $\frac{\sin\varepsilon-\varepsilon}{\varepsilon^3}$ is bounded as $\varepsilon \to 0$. Similarly we define $\sin\varepsilon \sim \varepsilon + o(\varepsilon^2)$ as $\varepsilon \to 0$ to mean $\frac{\sin\varepsilon-\varepsilon}{\varepsilon^2} \to 0$ as $\varepsilon \to 0$. In the second example (A.2), we could, with trepidation, expand the denominator of the integral by the binomial theorem and integrate formally to write

$$\int_0^\infty \frac{e^{-t}dt}{1+\varepsilon t} \overset{?}{\sim} \sum_{n=0}^{\infty}(-1)^n.n!\varepsilon^n \tag{A.3}$$

but the series on the right-hand side diverges for *all* $\varepsilon > 0$!

The key step is to have the courage not to discard this computation but rather exploit it by noting that when (A.2) is computed (say, using Simpson's rule), and compared with the first few (say N) terms of the series in (A.3) when ε is small, the results are in good agreement and get better and better as ε decreases for *fixed* N. What we must *not* do is let $N \to \infty$ as we would in the theory of convergent series. In fact it can be shown that

$$\int_0^\infty \frac{e^{-t}dt}{1+\varepsilon t} - \sum_0^N (-1)^n n! \varepsilon^n = O(\varepsilon^{N+1}) \quad \text{as} \quad \varepsilon \to 0 \quad \textit{for N fixed}$$

whereas

$$\sin \varepsilon - \sum_0^N \frac{(-1)^{n+1}\varepsilon^{2n+1}}{(2n+1)!} = \frac{O(\varepsilon^{2N+3})}{(2N+3)!} \quad \text{as} \quad N \to \infty \quad \textit{for fixed } \varepsilon$$

although we can *also* use this latter estimate as $\varepsilon \to 0$.

We can extend this idea both to expansions in terms of functions of ε other than powers and by allowing the coefficients to depend on a variable x. We first need to define an *asymptotic sequence* $\{\beta_n(\varepsilon)\}$ which consists of functions $\beta_n(\varepsilon)$ which decrease as $\varepsilon \to 0$ so that $\left|\frac{\beta_{n+1}(\varepsilon)}{\beta_n(\varepsilon)}\right| \to 0$ as $\varepsilon \to 0$. Then writing $f(x, \varepsilon)$ as an asymptotic expansion

$$f(x, \varepsilon) \sim \sum_0^\infty \alpha_n(x)\beta_n(\varepsilon) \tag{A.4}$$

means

$$f(x, \varepsilon) - \sum_0^N \alpha_n(x)\beta_n(\varepsilon) = O(\beta_{N+1}(\varepsilon)) \quad \text{as} \quad \varepsilon \to 0 \quad \text{for } \textit{N fixed.}$$

We again emphasise that we will never take an infinite number of terms on the right-hand side of (A.4), despite the presence of the ∞ symbol. Concerning the values of N which may be of most use to us, it is an interesting calculation on a pocket calculator to show that the error made when we go to N terms in (A.3) is smallest when ε is about N^{-1}: it increases rapidly if we take $\varepsilon > \frac{1}{N}$ and this "optimal error" is a characteristic of many asymptotic expansions. Indeed, it is possible to speculate further on the possibility of improving the accuracy by 'rearranging' the terms in the series corresponding to $n > \frac{1}{\varepsilon}$. In some cases these terms can be written as an integral for which a different asymptotic expansion is available and this idea is the basis of some modern theories of asymptotics (Olver).

Examples of what are hoped to be asymptotic expansions have appeared throughout these notes but only in a few of the "paradigms" (e.g., § 2.1 and 3.1) would it have been possible to prove that they satisfied the criterion (A.4). It is hoped that results such as (3.22) are in this category, but this may never be proved.

Whereas the above ideas were developed at the turn of the century by Poincaré, the key "modern" advance originated with the study of boundary layers in viscous flow. This was the invention of matched asymptotic expansions (m.a.e.'s) by Kaplun around 1950 (see Kevorkian and Cole). He realised that, when a function of x is expanded asymptotically for small ε, it will often happen that the expansion will be nonuniform, in that different α_n and β_n will be needed for different ranges of x (dependent on ε). However, he saw, crucially, that these different expansions will almost always coincide in what are called overlap regions of x. You can see this for yourself by scribbling down almost any function of x and ε (even $\frac{1}{x+\varepsilon}$) and expanding for small ε. When you have done this enough times you will also see the likely (but not universal) validity of the "matching principle" which states that if there are two regions for x (inner and outer) and the N term expansion in the inner region is (1) written in terms of the outer x variable and (2) expanded to M terms in ε, then the result is the same as if we had taken the M term expansion in the outer region and (3) written it in terms of the inner variable and finally (4) expanded to N terms in ε (see Van Dyke (1)). This may seem a mouthful and it needs modifying when logarithmic terms are present but it has proved the key to understanding countless problems both within and outside fluid dynamics.

Concerning the distinction between regular and singular expansions of the solutions of differential equations with independent variable x, a *regular* expansion is one in which the solution can be developed as a formal expansion in $\beta_n(\varepsilon)$, (usually with $\beta_n(\varepsilon) = \varepsilon^n$), in which the terms are found as usual by equating successive terms in β_n. A *singular* expansion is one in which rescaling is necessary to cater for different regions of x and this is more or less inevitable when ε multiplies the highest derivative in the differential equation. It is much harder to solve such problems because of the necessity of transferring information across the overlap domains, and this is where the matching principle is so useful.

We conclude by mentioning that the theory of asymptotic expansions is still wide open. This is because, if the β_n involve algebraic terms in ε, $f(x,\varepsilon)$ and $f(x,\varepsilon) + e^{-x/\varepsilon}$ and countless[1] other functions all have the *same* asymptotic expansion as $\varepsilon \to 0$. The problem of resolving the exponentially small terms at the "end" of an algebraic expansion occurs, for example, when computing the drag on a cylinder as Re $\to 0$, and it needs a more refined mathematical tool than anything described here. Further details and many examples can be found in Van Dyke (1), Bender and Orzag, Kevorkian and Cole, or Hinch.

[1] There are many more orders of magnitude (e.g., ε, ε^2, $e^{-1/\varepsilon}$, e^{-1/ε^2} ...) than there are real numbers: in fact they form a basis for the surreal numbers (Knuth).

Appendix B

Uses of group theory

Two of the most useful "tricks" used in these notes have been the lowering in order of certain ordinary differential equations (such as (2.21)) and the similarity reduction of certain partial differential equation problems to ordinary differential equation problems as in (2.3), (2.8), and (2.18). Even today it is not possible to give a watertight rationale for these mysterious occurrences but some insight can be gained from simple ideas of group theory; indeed, the study of differential equations was a great stimulus for the initiation of this subject in the nineteenth century.

Suppose we subject a differential equation (ordinary or partial) to an arbitrary change of variables (dependent, independent or both). Unless we are very lucky, a single such transformation will do us no good, but we have more chance of making progress when we consider *one-parameter* families of transformations.

B.1 Ordinary differential equations

To take a very simple example, suppose we know that the equation

$$\frac{dy}{dx} = F(x, y) \tag{B.1}$$

is invariant under the transformation $x' = x + \lambda$, i.e., $\frac{dy}{dx'} = F(x', y)$ for all λ. Then (B.1) is autonomous (i.e., $F(x, y)$ is independent of x) and the equation can be solved by separation of variables. Similarly we see that if any differential equation of order n is invariant under the transformation $x' = x + \lambda$ it can be reduced to a differential equation of order $n - 1$. We can extend this idea to the case where the equation is invariant under the transformation $g(x') = g(x) + \alpha(\lambda)$ where $g(x)$ and $\alpha(\lambda)$ are given functions by rewriting the equation with $g(x)$ as the independent variable. Thus the key question is: "Given a family of transformations $x' = \phi(x, \lambda)$ that leave a differential equation invariant, what characterises those ϕ for which there exists a change of variable $s = g(x)$ such that $s' = g(x') = s + \alpha(\lambda)$"? One answer to the question is that ϕ should form a *group* with respect to λ (called a Lie group, after its inventor). By this we mean that one value of λ (say $\lambda = 0$) must give the identity, there must be an inverse and associativity, but, most important, the group must be closed i.e. $x' = \phi(x, \lambda)$, $x'' = \phi(x', \mu)$ implies the existence of a $\nu(\mu, \lambda)$ such that $x'' = \phi(x, \nu)$. Every pair of values of the independent variable are related by the binary operation ϕ, and it is very easy to see that most functions of two variables which one writes down at random do *not* form a group with respect to their second argument.

To justify the above assertions[1] we first note that for ϕ to form a group it must satisfy the very strong constraint that $\frac{\partial \phi}{\partial \lambda}$ is 'more or less' a function just of ϕ. In fact, if we differentiate

$$\phi(\phi(x, \lambda), \mu) = \phi(x, \nu)$$

with respect to μ and set $\mu = 0$ we see that

$$\phi_2(\phi(x, \lambda), 0) = \phi_2(x, \lambda) \left.\frac{\partial \nu}{\partial \mu}\right|_{\mu=0}$$

where subscript 2 refers to differentiation with respect to the second argument, and hence $\frac{\partial \phi}{\partial \lambda}$ is a function of λ multiplied by a function of ϕ.

The dividend of having such a strong constraint on ϕ becomes clear when we now ask what happens to an arbitrary function $f(x)$ when we change variables. Remembering that $\phi(x, 0) = x$,

$$\begin{aligned} f(x') = f(\phi(x, \lambda)) &= f\left(x + \lambda \left.\frac{\partial \phi}{\partial \lambda}\right|_{\lambda=0} + \frac{\lambda^2}{2!} \left.\frac{\partial^2 \phi}{\partial \lambda^2}\right|_{\lambda=0} + \ldots\right) \\ &= f(x) + \lambda \left.\frac{\partial \phi}{\partial \lambda}\right|_{\lambda=0} f'(x) + \ldots \end{aligned}$$

and the tedium of the function of a function expansion collapses into, in shorthand,

$$f(x') = e^{\lambda U} f(x)$$

where $U = \left.\frac{\partial \phi}{\partial \lambda}\right|_{\lambda=0} \frac{d}{dx}$ is called the *infinitesimal generator* of the group. Hence, if we can solve $Ug = 1$, we can find our desired change of variable

$$s = g(x)$$

[1] For a more detailed account and geometric interpretation, see Bluman and Cole.

such that

$$s' = g(x') = g(x) + \alpha(\lambda).$$

You can follow this through in the simple case $\phi = (\lambda + 1)x$ to see that $s = \log x$, $\alpha = \log(1 + \lambda)$.

The above ideas can be extended to several variables, and in particular for two variables, the condition for $x' = \phi(x, y, \lambda)$, $y' = \psi(x, y, \lambda)$ to form a group is that $\frac{\partial}{\partial\lambda}\begin{pmatrix}\phi\\ \psi\end{pmatrix}$ should be a vector function of ϕ, ψ and U is now the partial differential operator $\left(\frac{\partial\phi}{\partial\lambda}\frac{\partial}{\partial x} + \frac{\partial\psi}{\partial\lambda}\frac{\partial}{\partial y}\right)_{\lambda=0}$. Now we need to be able to change from variables x, y to $r(x, y), s(x, y)$ where $Ur = 0$ and $Us = 1$ so that the group transformation is $r' = r$, $s' = s + \lambda$ and the equation in terms of $r(s)$ will be autonomous.

In the case of Blasius' equation (2.20), which is already autonomous, the first step is to write $f = X$, $f' = Y$ to transform it to

$$Y\frac{d^2Y}{dX^2} + \left(\frac{dY}{dX}\right)^2 + \frac{1}{2}X\frac{dY}{dX} = 0.$$

This equation is invariant under

$$X' = (1 + \lambda)X, \quad Y' = (1 + \lambda)^2 Y$$

and so $U = X\frac{\partial}{\partial X} + 2Y\frac{\partial}{\partial Y}$ is the appropriate infinitesimal generator. Then $Ur = 0$ and $Us = 1$ implies that $r = \frac{Y}{X^2}$ and $s = \log X$ and finally the equation becomes a second-order autonomous equation for $r(s)$ which can be transformed to a first-order equation for $\frac{dr}{ds}$ as a function of r (Chapter 2, exercise 12).

We can be even more general and consider transformations of x, y and $\frac{dy}{dx}, \frac{d^2y}{dx^2} \ldots$ (this leads to the idea of "jet spaces"). But, as stated earlier, such considerations have yet to produce anything like a complete theory. Indeed, a cynic might say that we have only been able to replace the task "spotting the transformation" by the task of "spotting the group", but there is no doubt that the latter task is much easier than the former. Nonetheless, the fact is that the theory has still not been able to specify those functions F which permit the solution of (B.1) by quadrature (i.e., indefinite integration).

B.2 Partial differential equations

The same idea can be applied to partial differential equations with the sole proviso that if boundary or initial conditions are to be satisfied, this will impose extra constraints on groups of transformations which leave the problem invariant. The principle remains the same: If we have a second-order equation in which x, y are independent variables and it is invariant under a group $x' = \phi(x, y, \lambda)$, $y' = \psi(x, y, \lambda)$ with infinitesimal generator U, then, when we change to variables r, s such that $Ur = 0$, $Us = 1$, the equation will be autonomous with respect to s and hence has solutions depending on r alone. For example, for (2.8) we note that the equation and boundary and initial conditions are invariant under $X' = (\lambda+1)^2 x$, $Y' = (\lambda+1)Y$ and we find that $r = Y/\sqrt{x}$ is a possibility. However, in honesty, we must mention that this particular example could have been explained immediately we had seen that if $T(x, Y)$ satisfied the problem, so did $T(\lambda^2 x, \lambda Y)$ for any real number λ. Then all we have to do is set $\lambda = \frac{1}{\sqrt{x}}$ for any fixed number x to see that T can only be a function of $\frac{Y}{\sqrt{x}}$!

A simple example in the theory of viscous flow shows that this theory is far from universal. If the fluid in $z > 0$ is at rest at infinity while the plane $z = 0$ rotates with constant angular velocity Ω, it is straightforward to see that there is a similar solution of the steady Navier-Stokes equations in the form

$$u_r = rF(z), \quad u_\theta = rG(z), \quad u_z = H(z)$$

where F, G and H satisfy certain ordinary differential equations (Chapter 1, exercise 19). In this case, however, there is no simple group invariance property of the kind described above.

We conclude by mentioning that it would be nice to be able to relate the ideas behind Lie groups, as described very crudely above, to those of their historical predecessors, Galois groups, which are so important in discussing the solvability of polynomial equations. Unfortunately Galois group theory depends crucially on several substantial auxiliary constructions and it would be inappropriate to describe it here. However, the basic idea is to see how certain algebraic functions of the roots transform under permutations of the roots (Dehn). The philosophy of seeing how the solutions of polynomial equations behave under permutation groups is very similar to that of seeing how the solutions of differential equations behave under continuous groups of transformations.

References

Acheson, D.J., 1990. *Elementary Fluid Dynamics*, Oxford University Press.
Batchelor, G.K., 1967. *An Introduction to Fluid Dynamics*, Cambridge University Press.
Bender, C.M., and Orzag, S.A., 1978. *Advanced Mathematical Methods for Scientists and Engineers*, McGraw-Hill.
Birkhoff, G., and Zarantonello, E.H., 1957. *Jets, Wakes and Cavities*, Academic Press.
Bluman, G., and Cole, J.D., 1974. "Similarity Methods for Differential Equations", *Appl. Math. Sci.* **13**, Springer-Verlag.
Buckmaster, J.D., and Ludford, G.S.S., 1982. *Theory of Laminar Flames*, Cambridge University Press.
Bush, A.W., 1992. *Perturbation Methods for Engineers and Scientists*, CRC Press.
Carslaw, H.S., and Jaeger, J.C., 1959. *Conduction of Heat in Solids*, Oxford University Press.
Chapman, S., and Cowling, T.G., 1952. *The Mathematical Theory of Nonuniform Gases*, Cambridge University Press.
Dehn, E., 1930. Algebraic Equations, Columbia University Press.
Fornberg, B., 1994 "Computing Steady Incompressible Flows Past Blunt Bodies – A Historical Overview" in *Numerical Methods for Fluid Dynamics IV* ed M.J. Barnes and K.W. Morton. Oxford University Press.
Greenspan, H.P., 1968. *The Theory of Rotating Fluids*, Cambridge University Press.
Hinch, E.J., 1991. *Perturbation Methods*, Cambridge University Press.
Kevorkian, J., and Cole, J.D., 1968. "Perturbation Methods in Applied Mathematics", *Appl. Math. Sci.* **34**, Springer-Verlag.
Knuth, D., 1974. *Surreal Numbers*, Addison-Wesley.
Langlois, W.E., 1964. *Slow Viscous Flow*, Macmillan.
Landau, L.D., and Lifshitz, E.M., 1987. *Fluid Mechanics*, 2nd ed., Pergamon.
Lighthill, M.J., 1978. *Waves in Fluids*, Cambridge University Press.
Milne-Thomson, L.M., 1948. *Theoretical Aerodynamics*, Macmillan.
Ockendon, H., and Ockendon, J.R., 1990. "Some thoughts on Macroscopic and Microscopic Mathematical Problems", *Math. Scientist* **15**, 15–23.
Olver, F.W.J., 1974. *Asymptotics and Special Functions*, Academic Press.
Proudman, I., and Pearson, J.R.A., 1957. "Expansions at Small Reynolds Numbers for the Flow Past a Sphere and a Circular Cylinder", *J. Fluid Mech.* **2**, 237–62.

Richardson, S., 1990. "How Not to Tackle Some Singular Perturbation Problems", *SIAM Review* **32**, 471–3.
Rosenhead, L., ed., 1963. Laminar Boundary Layers, Oxford University Press.
Saffman, P.G., 1993. *Vortex Dynamics*, Cambridge University Press.
Smith, F.T., 1982. "On the High Reynolds Number Theory of Laminar Flows", *IMA J. Appl. Math.* **28**, 207–81.
Stoker, J.J., 1957. *Water Waves*, Interscience.
Van Dyke, M. (1), 1975. *Perturbation Methods in Fluid Mechanics*, Parabolic Press.
Van Dyke, M. (2), 1982. *An Album of Fluid Motion*, Parabolic Press.
Whitehead, A.N., 1889. "Second Approximation to Viscous Fluid Motion", *Quart. J. Math.* **23**, 143–52.
Yih, C.-S., 1980. *Stratified Flows*, Academic Press.

Index